Modelling, state observation and control of simulated counterflow chromatographic separations

Vom Promotionsausschuss der
Fakultät für Elektrotechnik und Informationstechnik
der
Ruhr-Universität Bochum
zur Erlangung des akademischen Grades eines
Doktor–Ingenieurs
genehmigte Dissertation

von

Tobias Kleinert
aus Herdecke

2006

1. Gutachter: Prof. Dr.–Ing. J. Lunze,
Ruhr–Universität Bochum

2. Gutachter: Prof. Dr.–Ing. G. Brunner,
Technische Universität Hamburg–Harburg

Tag der mündlichen Prüfung: 25.11.2005

Bibliografische Information Der Deutschen Bibliothek

Die Deutsche Bibliothek verzeichnet diese Publikation in der Deutschen Nationalbibliografie; detaillierte bibliografische Daten sind im Internet über http://dnb.ddb.de abrufbar.

ISBN 3-8325-1222-5

Logos Verlag Berlin
Comeniushof, Gubener Str. 47,
10243 Berlin
Tel.: +49 030 42 85 10 90
Fax: +49 030 42 85 10 92
INTERNET: http://www.logos-verlag.de

Acknowledgements

This PhD thesis presents the results of my research work on the control of simulated counterflow chromatographic separations. Because of the complexity of the processes the design of control and observation algorithms, which are easy to parametrise and to implement, is a challenging task. This work contributes to the solution of this problem. It was carried out at the control institutes of Prof. Jan Lunze in Hamburg–Harburg and Bochum. The work was supported by Technische Universität Hamburg–Harburg and Ruhr–Universität Bochum.

I would like to express my deepest thank to Prof. Lunze for his excellent supervision. His deep insight into control theory and engineering as well as his courage to following new and non–standard ideas was a great inspiration for me and significantly influenced my research work.

I am very much obliged to Prof. Brunner for the opportunity to colaborate with his scientific coworkers and to perform experimental work at the Institute of Thermal Separation Sciences in Hamburg–Harburg, and for appraising the thesis. Many thanks go to Prof. Thomas Kriecherbauer from Bochum for the friendly support regarding the explicit solution of the true moving bed chromatography model. I would like to express my gratitude to Prof. Fischer, Prof. Awakowicz and Prof. Schwenk for their consent to be in my committee.

Of course, a deep thank goes to all my colleagues in Harburg–Harburg and Bochum. I would especially like to thank Thomas Steffen, Peerasaan Supavatanakul and Stanley Irungu Kamau. The fruitful scientific discussions and the mutual everyday support was a great help for me. Special thanks go to Monika Johannsen, Stephanie Peper and Matthias Lübbert, who supported me at Prof. Brunner's institute with respect to experimental work and technical discussions on chromatography. Many thanks go to Herwig Meyer for the design and implementation of an SMB flow controller, and to Stephanie, Monika, Matthias, Jan, Axel, Philippe, Jörg, Carsten, and Piotr for proof reading the manuscript. Deep thanks go to my students Birgit Koeppen, Randolph Ibe, Khalil Hannan, Ziqiang Wang, Mario Weiß, Jens Pache, Thomas Drescher and Jochen Blömken, who supported my work by their student and diploma projects.

I am grateful to my friends and fellow musicians for the wonderfull time and cooperation in Hamburg and the Ruhrgebiet. My family, especially my father, my sister and my brother and their families as well as my swiss relatives always gave me a warm wellcome in any occasion. They often simplified things for me during the realisation of this work.

Das Aufdecken von Zusammenhängen bedient die Neugier und inspiriert zum Entwickeln neuer Verfahren. Das ist meine Motivation zu forschen.

Mannheim, February 2006 Tobias Kleinert

Contents

Outline

Abstract

Simulated counterflow chromatographic separation plants consist of several chromatographic separation columns, which are connected to a circle by capillaries or tubes. A fluid circulates in the circle of columns and transports the fractions of a binary educt mixture. The component distributions propagate through the columns and are described by concentration profiles, which have the form of waves.

The inlet and outlet ports of the mixture and the products are attached to preselected column interconnections. Because of the propagation of the concentration profiles, the ports have to be permanently switched to new column interconnections at discrete switching times to limit the impurity of the products to a prescribed maximum.

These kinds of separation processes show a complex dynamical behaviour. Only few measurements of the concentration profiles are available. Because of parameter uncertainties, the exact offline determination of the operation conditions for a desired product purity is impossible. The separation underlies disturbances, which are e.g. the variation of the educt mixture concentration and the porosity of the adsorbent package.

This PhD thesis presents two new concepts for the control of the separation processes. The underlying goal is to derive control algorithms of low complexity.

The first concept considers time–invariant switching patterns of the inlet and outlet ports. The flow rates of the circulating fluids are adapted by decentralised discrete–time PI controllers such that the desired product purity is obtained. The concept uses selected, discrete–time concentration and purity measurements. For the reconstruction of the concentration profiles, a new approach is presented based on the explicit functional description of the wave front shape and propagation.

The control input variables of the second control concept are, in addition to the flow rates of the circulating fluids, the switching instants of the inlet and outlet ports. A rule–based discrete–event controller is designed, which determines the switching instants of the ports based on selected continuous–time concentration measurements of the wave fronts. The discrete controller is combined with a continuous controller to obtain the desired product purity.

The design of the controllers and the analysis of the closed–loop behaviour are performed based on numerical simulations using a detailed physical process model and experimental tests.

Deutsche Kurzfassung (German extended abstract)

Chromatographische Trennung mit simuliertem Gegenstrom

Chromatographische Trennanlagen mit simuliertem Gegenstrom bestehen aus ringförmig verschalteten chromatographischen Trennsäulen. In dem Ring zirkulieren Lösungsmittelströme, die die Komponenten A und B des binären Trenngemischs transportieren. Die Verteilung der Gemischkomponenten in den Säulen wird durch wandernde, wellenfrörmige Konzentrationsprofile c_A und c_B beschrieben (Bild 1).

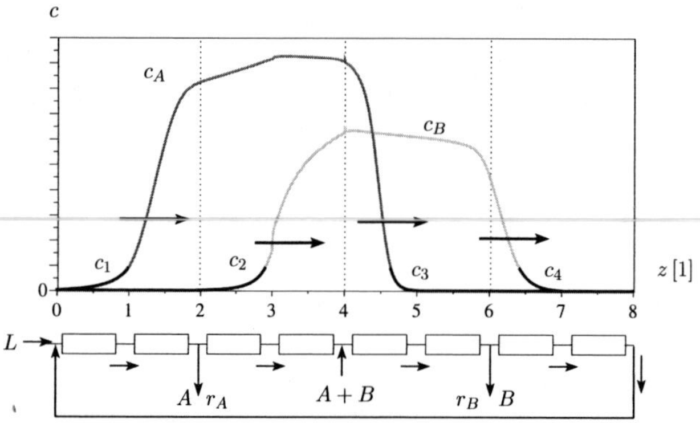

Bild 1: Wandernde Konzentrationsprofile und Wellenfronten in einem Ring von chromatographischen Trennsäulen

Die Zu- und Abflüsse des Gemisches ($A+B$), der Produkte (A, B) und des reinen Lösungsmittels (L) werden so zwischen den Trennsäulen platziert, dass die Verunreinigung der Produktströme durch die Wellenfronten c_1, c_2, c_3 und c_4 einen gegebenen Maximalwert nicht überschreitet. Durch diesen Vorgang ergeben sich die Produktreinheiten r_A und r_B. Aufgrund der Zirkulation der Konzentrationsprofile durch die Trennsäulen müssen die Zu- und Abflüsse permanent neu positioniert, d.h. geschaltet, werden. Es wird zwischen zwei Fahrweisen unterschieden: Das synchrone Schalten aller Zu- und Abflüsse wird als Simulated–Moving–Bed–Prinzip (SMB) und das asynchrone Schalten als Variable–Length–Column–Prinzip (VARICOL) bezeichnet. Die Schaltfolge der Zu– und Abflüsse wird durch ein Schaltmuster bestimmt, das fest vorgegeben sein kann und daher zeitlich invariant ist, oder von einer Steuerung auf der Basis des Anlagenzustand bestimmt wird und somit zeitvariabel sein kann.

Die Trennprozesse haben ein komplexes dynamisches Verhalten und es stehen nur wenige Messgrößen zur Verfügung. Wegen Parameterunsicherheiten kann der Arbeitspunkt für die spezifizierten Produktreinheiten nicht exakt bestimmt werden. Die Trennung wird z.B. durch variierende Eintrittskonzentrationen und Adsorbensporositäten gestört.

Beobachtungs– und Regelungsansätze

SMB– und VARICOL–Prozesse werden durch einen Satz nichtlinearer partieller Differentialgleichungen mit schaltenden Anfangs– und Randbedingungen und schaltenden Modellparametern beschrieben. Aufgrund der erheblichen Modellkomplexität und der nur ungenau bekannten physikalischen Parameter führt der Entwurf von Zustandsbeobachtern und Reglern auf der Grundlage des physikalischen Modells zu sehr komplexen Algorithmen, die nur bedingt praxistauglich sind. Es muss berücksichtigt werden, dass nur wenige Konzentrationsmesswerte verfügbar sind.

Im Rahmen dieser Dissertation wird die Modellreduktion von SMB– und VARICOL–Prozessen behandelt, die für den Entwurf einer Zustandsbeobachtung der Wellenfronten des SMB–Prozesses und den Entwurf dezentraler, zeitdiskreter Regler der Wellenfronten und Produktreinheiten von SMB– und VARICOL–Prozessen genutzt wird. Des weiteren wird eine Steuerung von VARICOL–Prozessen vorgestellt, die sowohl die Umschaltzeitpunkte der Zu– und Abflüsse als auch die Flussraten der zirkulierenden Fluide anpasst. Das zentrale Ziel ist die Ableitung von Beobachtungs- und Steuerungsalgorithmen, die eine geringe Komplexität aufweisen, leicht in einem Prozessführungssystem implementiert werden können und deren Parameter einfach zu bestimmen sind.

Prozessbeschreibung und detaillierte Modellierung

Mit Bezug auf die zentralen Themen der Regelung und Steuerung werden SMB– und VARICOL–Prozesse und deren Betriebsweise detailliert erklärt. Das Konzept der Wellenfronten und der zeitdiskreten, direkten Messung von ausgewählten Konzentrationen und der Produktreinheiten wird dargestellt. Auf der Grundlage der detaillierten Modellierung des diskreten und kontinuierlichen Teilsystems wird das hybride dynamische Verhalten von SMB– und VARICOL–Prozessen mittels hybrider Automaten beschrieben.

Rekonstruktion der Wellenfronten

Für die Rekonstruktion der Wellenfronten wird ein stark vereinfachtes Modell abgeleitet, das

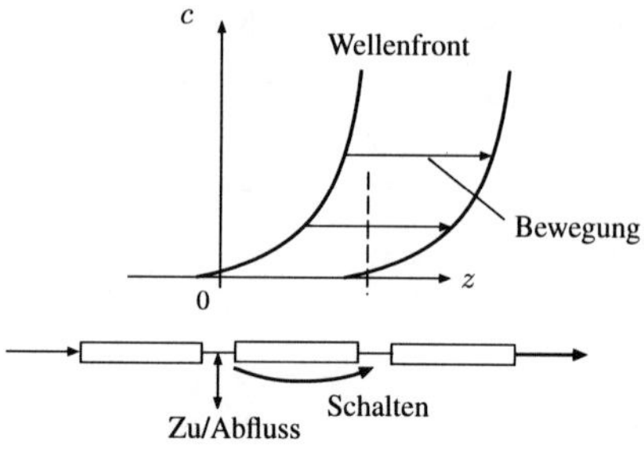

Bild 2: SMB–Wellenfront

die Form, die Position und die Bewegung der Wellenfronten der SMB–Konzentrationsprofile mit einem Formansatz beschreibt (Bild 2). Der explizite, funktionale Formansatz wird aus der stationären Lösung des wahren Gegenstromprozessmodelles abgeleitet. Die drei Parameter der funktionalen Beschreibung werden als Zustandsvariable der Wellenfront interpretiert. Auf der Grundlage eines zeitdiskreten, zeitvariablen Zustandsraummodells der Parameter wird mit einem zeitvariablen Luenberger–Beobachter ein Algorithmus zur Rekonstruktion der Wellenfronten für den stationären Betrieb und den dynamischen Übergang, der aufgrund von Änderungen des Adsorptionsverhalten, der Festbettporosität und der internen Fluidströme auftritt, vorgestellt.

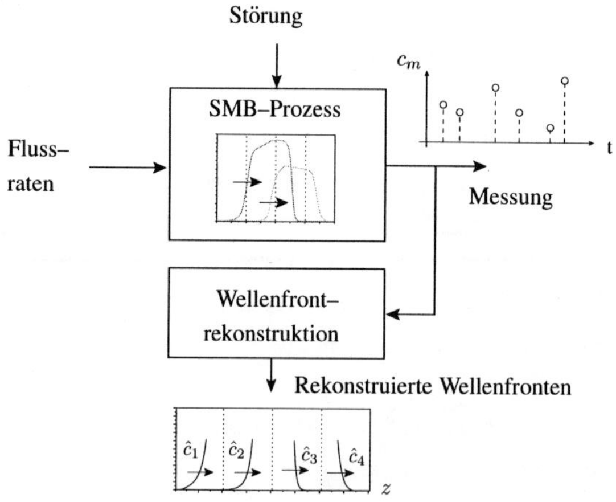

Bild 3: SMB–Wellenfrontrekonstruktion

Der Wellenfrontbeobachter basiert auf der zeitdiskreten Messung c_m der Wellenfrontkonzentrationen (Bild 3). Für die Auslegung ist lediglich die Kenntnis der Dauer der Umschaltperiode und des Messzeitpunktes erforderlich. Mit einer Untersuchung des Beobachterfehlers wird gezeigt, dass der Beobachter insbesondere bei SMB–Prozessen, die eine gute Übereinstimmung mit dem wahren Gegenstrom zeigen, sehr gute Ergebnisse liefert. Die Erprobung des Verfahrens in einer numerischen Simulation auf Basis des physikalischen Prozessmodells wird anhand der Trennung eines Tocopherolgemisches dargestellt.

Regelung bei zeitinvarianten Schaltmustern

Bezüglich der Regelung wird zwischen wertekontinuierlichen und wertediskreten Prozesseingangsgößen $\boldsymbol{u}$ und v unterschieden. Der Vektor $\boldsymbol{u}$ fasst die Flussraten der zirkulierenden Fluide zusammen. Der Vektor v bezeichnet die Schaltsignale der Zu– und Abflüsse. Die wertekontinuierlichen Störgrößen werden in $\boldsymbol{d}$ zusammengefasst. Es wird angenommen, dass als Ausgangsgröße zeitdiskret oder zeitkontinuierlich gemessene Konzentrationen der Gemischkomponenten in den Säulenverbindugen und die zeitdiskret gemessenen Produktreinheiten verfügbar sind.

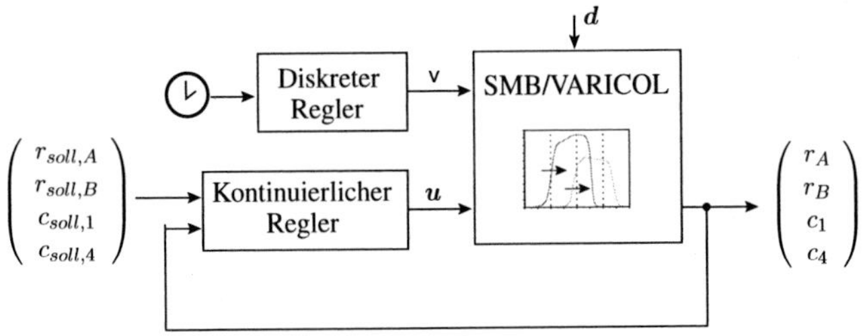

Bild 4: Regelung von SMB– und VARICOL–Prozessen mit zeitinvariantem Schaltmuster

Wenn SMB– und VARICOL–Prozesse mit zeitinvarianten Schaltmustern betrieben werden, wird v durch einen diskreten Regler in einer offenen Steuerkette vorgegeben. In diesem Fall können die Trennprozesse als zeitdiskrete, wertekontinuierliche Systeme beschrieben werden. Für diesen Ansatz werden im Rahmen dieser Dissertation zwei neue Regelungskonzepte vorgestellt und untersucht, die auf der zeitdiskreten Messung der Produktreinheiten beruhen:

1. Das erste Konzept basiert auf einer örtlich mitbewegten, zeitlich äquidistanten Abtastung der Wellenfronten c_1 und c_4 in der Säulenverbindung, in der das Lösungsmittel L zugeführt wird, und der zeitdiskreten Messung der Produktreinheiten. Die Messgrößen werden auf einen kontinuierlichen Regler zurückgeführt, der aus dezentralen, zeitdiskreten PI–Reglern besteht und die Flussraten der zirkulierenden Fluide so anpasst, dass die gewünschten Produktreinheiten erreicht werden (Bild 4). Das Eingangs–Ausgangs–Verhalten der Regel-

strecke kann mit vier entkoppelten, linearen Eingrößensystemen erster Ordnung approximativ beschrieben werden, was die Reglerauslegung stark vereinfacht. Die Modellidentifikation sowie die Erprobung des Regelverhaltens erfolgt mit numerischen Simulationen auf der Grundlage des physikalischen Prozessmodells. Das dynamische Verhalten wird anhand von Experimenten an einer konkreten Anlage bestätigt. Der Ansatz ist auf SMB– und VARICOL–Prozesse anwendbar.

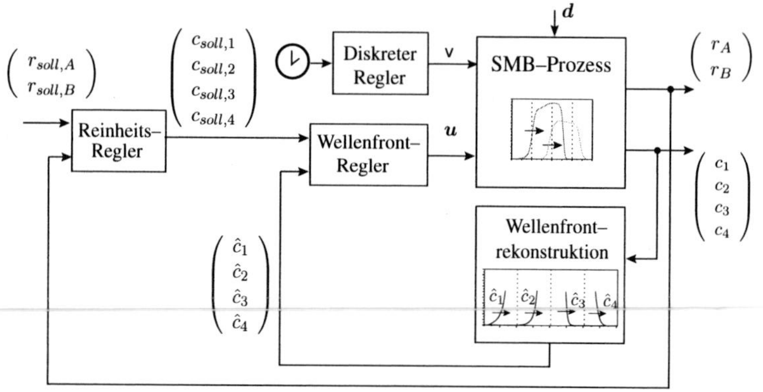

Bild 5: Beobachterbasierte Regelung von SMB–Prozessen

2. Bei steilen Wellenfronten führen kleine Änderungen in den kontinuierlichen Eingangsgrößen zu starken Änderungen in den Messgrößen. Dies erschwert die Auslegung der Regler. Um das Problem zu umgehen wird die Rekonstruktion der Wellenfronten eingesetzt, um Veränderungen frühzeitig erkennen zu können (Bild 5). Über die Rückführung der Information über die Form und Bewegung der Wellenfronten können deren Positionen mit einem Wellenfrontregler, bestehend aus dezentralen, zeitdiskreten PI–Reglern, geregelt werden. Mit der Rückführung der Produktreinheiten in einer äußeren Reglerkaskade werden die Sollwerte für die Wellenfrontpositionen vorgegeben. Wiederum kann das Eingangs–Ausgangsverhalten der Regelstrecke durch entkoppelte, lineare Eingrößensysteme erster Ordnung beschrieben werden. Da die Rekonstruktion der Wellenfronten nur für SMB–Prozesse abgeleitet wurde, wird dieser Ansatz an einer numerischen Simulation eines SMB–Prozesses erprobt. Prinzipiell ist die Übertragung auf VARICOL–Prozesse möglich.

Mit den Reglern ist das automatische Anfahren der Trennprozesse in einen vorgegebenen Arbeitspunkt und das Ausregeln von Störungen möglich. Dadurch, dass für die Auslegung der Regler nur wenige, leicht bestimmbare Parameter der Trennprozesses erforderlich sind, ergibt sich eine wesentliche Erneuerung im Vergleich zu bisher bekannten Verfahren, die auf physikalischen Prozessmodellen aufbauen. Zusätzlich sind die einfachen Regler bei der Implementierung leicht handhabbar.

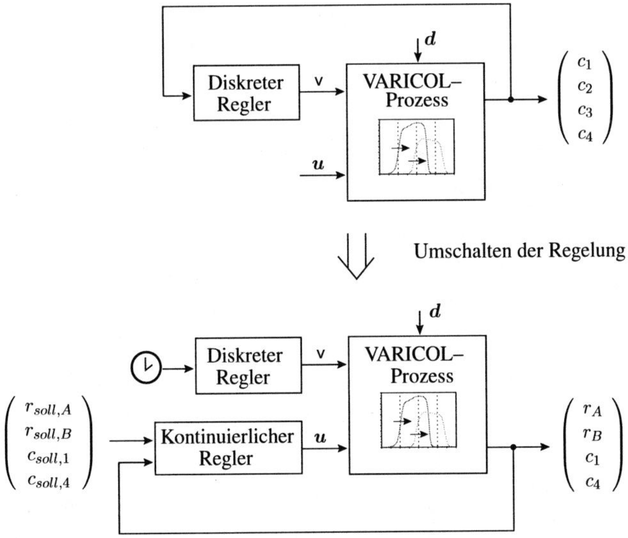

Bild 6: Kombinierte diskrete und kontinuierliche Regelung von VARICOL–Prozessen

Regelung bei zeitvariablen Schaltmustern

Geeignete Schaltmuster für VARICOL–Prozesse können wie die Arbeitspunkte für SMB–Prozesse nur ungenau im Voraus bestimmt werden. Daher wird ein diskreter, regelbasierter Regler vorgeschlagen, mit dem auf der Basis der kontinuierlichen Messung der Wellenfrontkonzentrationen die Schaltzeitpunkte der Zu- und Abflüsse bestimmt werden. Bei konstanten kontinuierlichen Eingangsgrößen überführt der diskrete Regler den Trennprozess in einen zyklisch stationären Zustand, für den der Regler ein geeignetes Schaltmuster vorgibt. Anschließend wird von der diskreten auf eine kontinuierliche Regelung umgeschaltet, wobei das stationäre Schaltmuster beibehalten und der Trennprozess in den gewünschten Arbeitspunkt überführt wird (Bild 6). Da das Schaltmuster nicht im Voraus bekannt ist, kann das dynamische Verhalten des kontinuierlich geregelten Systems nicht vorab bestimmt werden. Daher wird für die kontinuierliche

Regelung ein modifizierter Dreipunktregler eingesetzt, für den sich unabhängig vom geregelten Prozess eine Regelabweichung ergibt, die in einer vorgegebene Schranke liegt.

Ergebnisse

Diese Dissertation leistet einen Beitrag zur Modellierung, Zustandsbeobachtung und Regelung von SMB– und VARICOL–Prozessen. Zu den folgenden Themen werden neue Ergebnisse dargestellt:

- ◇ Analyse des Zusammenhangs zwischen den Wellenfronten und dem Betriebsverhalten von SMB– und VARICOL–Prozessen
- ◇ Direkte Messung der Produktreinheiten von SMB– und VARICOL–Prozessen
- ◇ Herleitung der Konvektions–Diffusions–Gleichung als physikalisches Modell der chromatographischen Stofftrennung mit wahrem Gegenstrom sowie analytische, explizite Lösung des stationären Konzentrationsverlaufs
- ◇ Modellierung der kontinuierlichen und diskreten Teilsysteme und Modellierung des Gesamtprozesses mit hybriden Automaten
- ◇ Modellierung und Rekonstruktion der Wellenfronten in SMB–Prozessen auf Basis eines Fromansatzes und eines linearen, zeitdiskreten Luenberger–Beobachters
- ◇ Dezentrale Regelung von SMB– und VARICOL–Prozessen mit zeitdiskreten PI–Regler bei zeitinvarianten Schaltmustern
- ◇ Kombinierte diskrete und kontinuierliche Regelung von VARICOL–Prozessen bei zeitvariablen Schaltmustern

Zur Analyse des Prozessverhaltens und zur Verifikation der Ergebnisse wurde die MATLAB–Toolbox CSep (für "Chromatographic Separation") entwickelt. Sie ermöglicht die numerische Simulation des dynamischen Anlagenverhaltens auf der Grundlage von physikalischen Prozessmodellen.

Die Herleitung reduzierter Prozessmodelle für die Rekonstruktion der Wellenfronten und für die Reglerauslegung basiert auf numerischen Simulationen und experimentellen Tests. Die Rekonstruktion und die Regelungsverfahren wurden anhand von numerischen Simulationen getestet.

Chapter 1

Introduction

Separation processes isolate single fractions of an educt mixture. Simulated Moving Bed (SMB) and Variable Length Column (VARICOL) processes implement the continuous chromatographic separation using a simulated counterflow of the solvent and the adsorbent. The process behaviour is described by the superposition of discrete and distributed continuous dynamics. Because of the complexity of the physical setup an automatic control of the units is desirable, but yields complex control laws if based on physical plant models. This chapter briefly describes the properties of SMB and VARICOL processes and the approaches to the control of these processes which are treated in this thesis. The aim and the main results are lined out. The related literature is discussed and the structure of the thesis is given at the end of the chapter.

1.1 Simulated Counterflow Chromatography

Separation processes are important intermediate steps in chemical production lines. Chromatographic separation techniques are applied in the oil and food industry in large scale productions of several hundred tons per year and have gained considerable importance in the field of life science and pharmaceutics. Simulated Moving Bed (SMB) and Variable Length Column (VARICOL) processes are continuous chromatographic separation processes, which simulate the counterflow between the adsorbent and the solvent. Within this thesis, these processes are referred to as Simulated Counterflow Chromatographic (SCC) processes.

The processes perform a continuous separation of binary mixtures $A+B$, with A and B denoting the mixture components. The mixture is continuously fed into the separation plant by the *feed flow* and the separated components A and B are recovered from the plant in the *product streams*. Inlet and outlet ports are used to inject the mixture and to withdraw the product streams, respectively. To compensate the solvent loss at the product outlets, pure solvent S is injected to the plant through the solvent inlet port.

The inlet and outlet ports are attached to the interconnections of a circle of separation columns

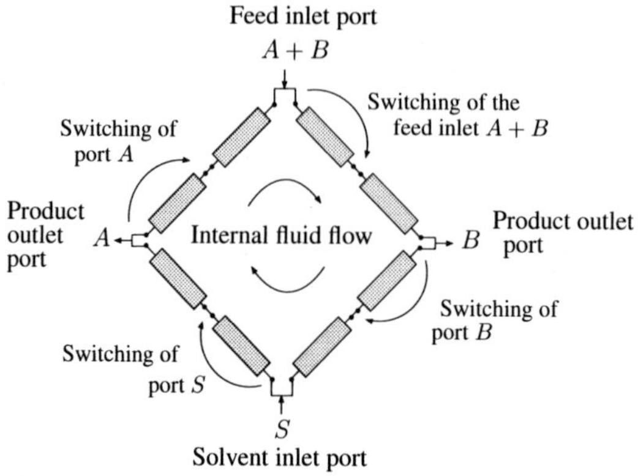

Figure 1.1: Inlet and outlet ports and port switching

(Figure 1.1). The positions of the inlets and outlets are changed stepwise by a unidirectional discrete switching. In SMB processes, all ports are switched *synchronously*, while in VARICOL processes *asynchronous* port switching is applied. The fluid, which circulates in the circle of the separation columns, causes the propagation of the distribution of the components A and B. These component distributions are described by *concentration profiles* c_A and c_B which have the form of bell curves or waves with respect to the spatial coordinate z and which propagate at a certain distance of each other through the separation columns (Figure 1.2). Due to the overlapping of c_A and c_B the by–product concentrations at the product outlets are not zero and, hence, impurities of the product streams occur. The spatial delimitation of the concentration profiles and thereby the by–product concentrations at the corresponding outlet ports are determined by the propagation of the *wave fronts* c_1, c_2, c_3 and c_4.

Process operation. To obtain satisfactory product purity values r_A and r_B, the switching of the ports has to be performed in such a way that, at the positions of the product outlet ports, the overlapping of the concentration profiles is low and high product concentrations are obtained. The fluid which circulates through the columns is determined by the internal fluid flow rates $\dot{m}_I$, $\dot{m}_{II}$, $\dot{m}_{III}$ and $\dot{m}_{IV}$ (Figure 1.2). Because these fluid flows influence the propagation of the concentration profiles with respect to the solvent ports, the operation mode of SCC processes is determined by the interaction between the *switching of the inlet and outlet ports* and the *adaptation of the internal fluid flow rates*. The port switching is a discrete action and is referred to as the discrete control input of SCC processes. The variation of the fluid flow rates is referred to as a continuous action and, therefore, the internal fluid flow rates are the continuous control input.

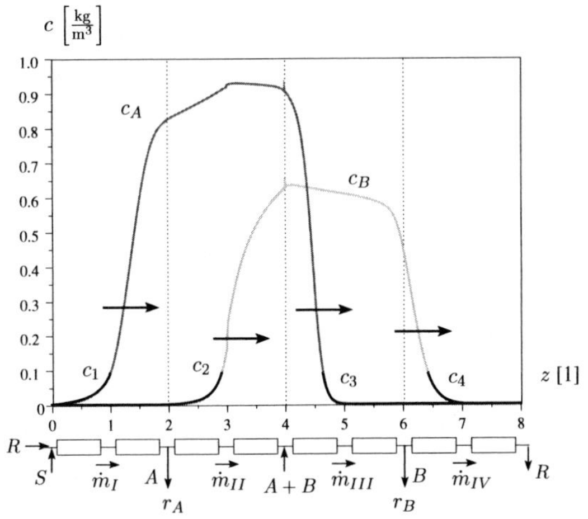

Figure 1.2: Concentration profiles and internal fluid flow rates in an SCC plant

The superposition of discrete port switching and continuous mass transport processes yields a complex dynamical behaviour. Because of uncertain physical parameters, it is not possible to precisely determine in advance the operation point which yields the desired product purity. SCC processes are influenced by disturbances, e.g. changes of the feed inlet concentration or the porosity of the adsorbent packages, or further effects, which change the shape and propagation of the concentration profiles.

Measurements. Because of the physical setup of the processes only few measurements of the concentration profiles and the purity values are available. However, it is possible to perform selected measurements of the concentrations c_A and c_B at the column interconnections, but with a low sampling rate. It is also possible to record continuous measurements of c_A and c_B by measurement units mounted in the column interconnections. Furthermore, it is possible to perform the discrete–time measurement of the product purity values r_A and r_B which occur in the time span between two switchings of the corresponding product outlet port. However, it is impossible to measure the complete concentration profiles c_A and c_B.

Necessity of process control. In industrial practice, SCC processes are open–loop controlled. Usually, a conservative operation point is chosen in the sense that there is a large security margin with respect to the desired purity values of the product streams. This guarantees a certain robustness against disturbances. To be able to drive the processes to the operation point which yields the desired product purity values, an automatic control is necessary. This requires, on the one hand, a controller design model of the dynamical input–output behaviour of SCC processes and, on the other hand, the measurement or the reconstruction of the controlled variables.

1.2 Control task and ways of solution

Control task. From the point of view of the controller design, the SCC process is a multiple–input multiple–output system (Figure 1.3). It has the discrete control input

$$\mathsf{v} = \begin{pmatrix} E_S \\ E_A \\ E_{A+B} \\ E_B \end{pmatrix},$$

where E_S, E_A, E_{A+B} and E_B are binary signals with the value 0 or 1 (which corresponds to "don't switch" or "switch", respectively) and the continuous control input

$$\boldsymbol{u} = \begin{pmatrix} \dot{m}_I \\ \dot{m}_{II} \\ \dot{m}_{III} \\ \dot{m}_{IV} \end{pmatrix}$$

of the internal fluid flow rates. Further input variables are the disturbance input

$$\boldsymbol{d} = \begin{pmatrix} c_{A+B,A} \\ c_{A+B,B} \\ \varepsilon \end{pmatrix},$$

where $c_{A+B,A}$ and $c_{A+B,B}$ are the feed inlet concentrations of the components A and B and ε is the porosity of the adsorbent package in the separation columns.

Two kinds of measured output variables are considered. The continuous–time output $\boldsymbol{y}_c$ compounds the continuously measured concentrations

$$\boldsymbol{y}_c = \begin{pmatrix} c_{c,A}(z_{m1}) \\ c_{c,B}(z_{m1}) \\ c_{c,A}(z_{m2}) \\ c_{c,B}(z_{m2}) \\ \vdots \end{pmatrix},$$

where the index c indicates the continuous–time measurement and z_{mi}, $i = 1, 2, \ldots$ are selected measurement position with respect to the spatial coordinate z. The discretely measured purity values and concentrations are represented by the output variable

$$\boldsymbol{y}_d = \begin{pmatrix} r_{d,A} \\ r_{d,B} \\ c_{d,A}(z_{m1}) \\ c_{d,B}(z_{m1}) \\ c_{d,A}(z_{m2}) \\ c_{d,B}(z_{m2}) \\ \vdots \end{pmatrix},$$

where the index d indicates that discrete–time measurements are considered. Figure 1.3 shows the block diagram of the SCC process.

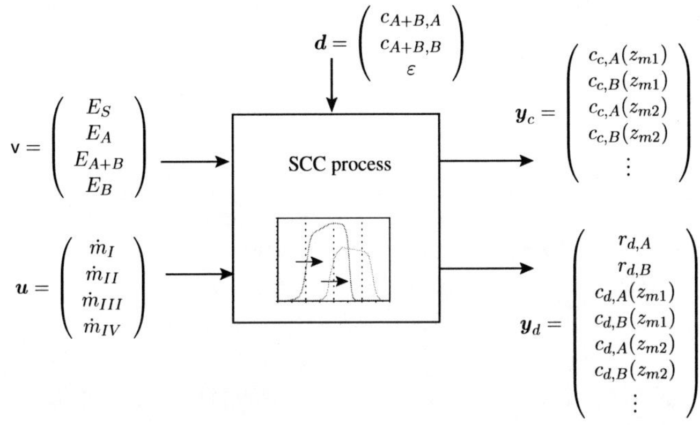

Figure 1.3: Input and output variables of SCC processes

The aim of SCC process control is to keep the product purity values $r_{d,A}$ and $r_{d,B}$ close to the set–point values $r_{set,A}$ and $r_{set,B}$. The dynamical behaviour of the SCC shows certain properties which allow for a significant reduction of the complexity of the controller design model compared to a detailed physical process model. This property is regarded for the design of a closed–loop SCC control. The following **control task** is considered:

Task 1.2.1 *Given are an SCC process according to Figure 1.3 and a conservative operation point which guarantees a high security margin of the purity values to the set–point. Find a control law which prescribes the switching of the inlet and outlet ports and the manipulation of the internal fluid flow rates such that the desired product purity values are obtained and kept in case of disturbances of the feed inlet concentrations and the adsorbent package porosity.* □

Ways of solution. For the solution of Task 1.2.1 it is investigated, how the impurity of the product streams can be manipulated in order to control the product purity. The impurity of the products is determined by the amount of the by–product at the considered product outlet. The by–product concentrations in the product streams are determined by the wave fronts of the respective by–product concentration profile. Hence, a controller has to manipulate the control input in such a way that the propagation of the wave fronts with respect to the product outlet ports yields the desired product purity values. Regarding the control input variables v and $\boldsymbol{u}$, there are three options for a control concept, which are independent of the considered separation plant setup or the chemical separation problem:

1. **Continuous control of SCC processes**: Perform a time–triggered switching which yields a repeated, time–invariant sequence of the port switchings, and manipulate the internal fluid flow rates $\dot{m}_I$, $\dot{m}_{II}$, $\dot{m}_{III}$ and $\dot{m}_{IV}$ such that the desired purity values are reached and held in case of disturbances.

2. **Discrete control of SCC processes**: Keep the internal fluid flow rates constant and perform a switching of the ports considering the movement of the wave fronts such that low by–product concentrations are obtained at the product outlets.

3. **Combined discrete and continuous control of SCC processes**: Combine the two previous approaches in the sense that the port switching leads to low by–product concentrations in the product streams, and the manipulation of the internal fluid flow rates yields the desired purity values.

1.3 Aims and main results of the thesis

This thesis presents new concepts for the continuous, the discrete and the combined discrete and continuous control of SCC processes, which solve Task 1.2.1. The underlying aim is to derive control laws of low complexity based on simple controller design models. The following paragraphs summarise the concepts and point out their novelty. Besides of these, the thesis presents new results on the modelling of the combined continuous and discrete dynamics of SCC processes and on the reconstruction of the concentration profiles.

Continuous control. If the switching signal v is generated by a discrete controller which is triggered by a clock (Figure 1.4), the sequence of switching signals forms a repeated, time–invariant pattern. A control concept is proposed in Chapter 5, which uses discrete–time purity measurements $r_{d,A}$ and $r_{d,B}$, and discrete–time wave front concentration measurements $c_{d,1}(z_m)$ and $c_{d,4}(z_m)$ which are recorded at the position z_m. The measurements are directly fed back to the continuous controller, which manipulates the continuous control input $\boldsymbol{u}$ such that the set–points $r_{set,A}$, $r_{set,B}$, $c_{set,1}$ and $c_{set,4}$ are reached and disturbances $\boldsymbol{d}$ are attenuated. For the derivation of a controller design model, the input–output behaviour is analysed using methods for multivariable systems (Lunze, 1989, 2004b; Raske, 1981). The dynamics can be approximated by four decoupled first–order singel–input single–output systems. Decentralised discrete–time PI controllers are applied to solve Task 1.2.1.

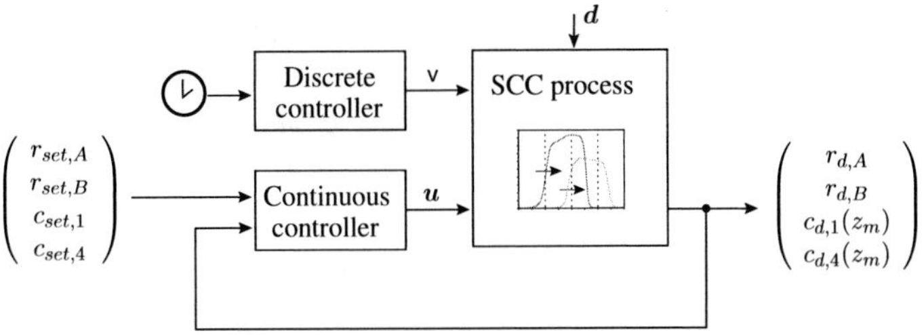

Figure 1.4: Continuous control of SCC processes

The concept is similar to the control of the SMB wave front positions presented in (Klatt et al., 2002) or the inferential control concept proposed in (Schramm et al., 2003). Both of these approaches use linear discrete–time controllers to stabilise the wave front positions. However, a cascaded controller has to be applied for the control of the product purity values. For the concept shown in Figure 1.4, no cascaded control is necessary. Furthermore, the controller tuning follows a simple procedure and the concept applies to SMB as well as to VARICOL processes.

In the following, this concept is referred to as the *measurement–based continuous control of SCC processes.*

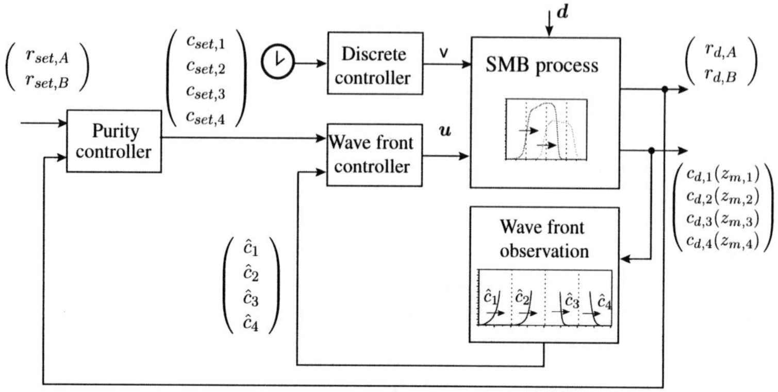

Figure 1.5: Observation–based control of SMB processes

For special separation problems with low diffusive effects and nonlinear adsorption, the reconstruction of the wave fronts c_1, c_2, c_3 and c_4 is necessary, which is referred to as the *wave front observation*. The new solution of this task which is presented in Chapter 4 uses an explicit functional description of the shape and propagation of each wave front. A simple third order discrete–time system describing the dynamical behaviour of the parameters of the functional descriptions is derived. Based on a discrete–time Luenberger observer a new concept for the wave front observation is derived using discrete–time concentration measurements $c_{d,i}(z_{m,i})$ at the positions $z_{m,i}$ which are specific for the wave fronts c_i, $i = 1, 2, 3, 4$. The observer output is fed back to a wave front controller which stabilises the wave front concentrations $\hat{c}_i$ to a given set–point $c_{set,i}$, $i = 1, 2, 3, 4$ (Figure 1.5). For the control of the purity values an outer control loop with the purity controller is necessary, which prescribes the wave front set–points based on the control offset of the discretely measured purity values $r_{d,A}$ and $r_{d,B}$. Both the wave front controller and the purity controller are decentralised discrete–time PI controllers (Chapter 5). This control concept is similar to the concept proposed in (Hanisch, 2002). The significant difference is the direct measurement of the product purity values. The novelty of this concept arises from the use of the new wave front observer within the cascaded controller scheme. Furthermore, the use of the concept is not restricted to the control of SMB processes, since it can principally be applied to VARICOL processes. This control concept is referred to as the *observation–based continuous control of SCC processes*.

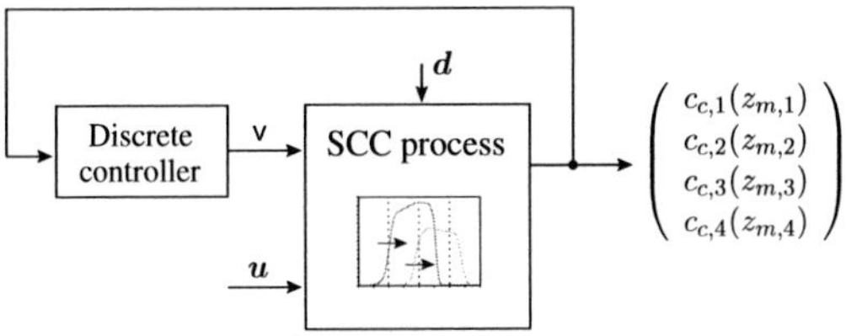

Figure 1.6: Discrete control of the VARICOL port switching instants

Discrete control. In an alternative control concept, the switching signal v is generated by discrete closed–loop control, based on the continuous measurement of the wave front concentrations $c_{c,i}$ at specific positions $z_{m,i}$, $i = 1, 2, 3, 4$ (Figure 1.6).
The discrete controller consists of a set of rules for the placement of the inlet and outlet ports with respect to the wave fronts in order to achieve a low impurity at the product outlets (Chapter 6). Because this concept leads to an asynchronous switching in the general case, it is referred to as the discrete control of VARICOL processes. The concept is the first known approach to the closed–loop control of the switching instants for the VARICOL process and is entirely based on concentration measurements. It is referred to as the *discrete control of VARICOL processes based on continuous wave front concentration measurements*.

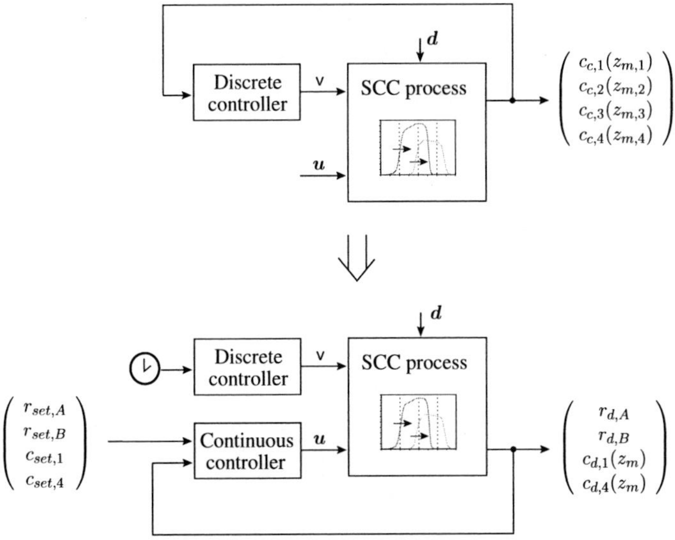

Figure 1.7: Combined discrete and continuous control

Combined discrete and continuous control. An unfavourable choice of the continuous control input u and the presence of disturbances might render the aim impossible to reach the set–point purity values by a discrete closed–loop control alone. In that case, a combination of discrete and continuous control is necessary. The following concept is proposed: The discrete closed–loop control is applied during the start–up of the plant. When the stationary state is reached, the resulting switching pattern is applied as a time–invariant switching pattern and a continuous controller is applied to drive the plant to the set–point (Figure 1.7). The continuous controller which is applied for this concept consists of decentralised modified three–step controllers (Chapter 6). This method allows for the closed–loop control of the VARICOL process using the discrete *and* the continuous control input variables. It is referred to as the *combined discrete and continuous control of VARICOL processes*.

Main results. For the solution of Task 1.2.1 three topics in the field of SCC control are concerned in this thesis. The following list shows the main results:

- A detailed **SCC process modelling and analysis**, considering the *distinct continuous and discrete dynamics* as well as the *interaction* between both, is performed.
  A physical model based on the Convection–Diffusion–Equation is derived for the description of the *continuous dynamics*, which allows to determine the influence of the disturbance and the continuous and the discrete control inputs onto the model equations.
  A general model of the *discrete dynamics* of SCC processes is derived in form of an automaton representation, including a state transition and output relation and the representation of the discrete state trajectory by automaton graphs. The concept is applied to the SMB and to VARICOL processes considering time–varying and time–invariant switching patterns.
  The interaction between the continuous and the discrete model is considered for the modelling of the *overall SCC process dynamics*. For the representation and the analysis of the overall system dynamics the concept of hybrid automata is used.
  Based on the detailed process modelling the interaction between the discrete and the continuous part of the SCC is analysed. The mixed continuous and discrete trajectory of the process is represented.

- With respect to the **reconstruction of the SMB concentration profiles**, a new concept is described which yields an *observation algorithm of low complexity*. For the design of the observer only measurement times, switching times and the observer error dynamics have to be provided. No further physical process parameters are necessary.

- With respect to the **control of SCC processes**, distinct approaches for time–invariant and time–varying switching patterns are presented. In both cases, strongly simplified controller design models are derived such that *control algorithms of low complexity* can be applied. Hence, despite of the complexity of the process, the presented solutions of Task 1.2.1 are considerably easy to handle with respect to a practical application.

1.4 Literature review

Process modelling and analysis. Regarding the detailed modelling of SCC processes, three topics have to be considered. The first refers to the modelling of the continuous SCC dynamics, which are governed by coupled fluid dynamical systems. Many publications can be found on this topic (Golshan-Shirazi and Guiochon, 1992; Seidel-Morgenstern, 1995; Dünnebier et al., 1998; Ruthven and Ching, 1998; Klatt, 1999; Dünnebier and Klatt, 2000; Spieker, 2000; Giese, 2002). Among the proposed models, the Equilibrium–Dispersive–Model, which encounters the Convection–Diffusion–Equation, is regarded as a suitable approach with respect to model accuracy and complexity, if low diffusive effects can be considered (Giese, 2002). It allows for a compact representation of the continuous dynamics regarding the distributed parameters and for effective numerical simulation (Kleinert, 2002).

The second topic refers to the discrete dynamics which is governed by deterministic discrete–event systems. Modelling concepts for this system class are described e.g. in (Lunze, 2003). Regarding SCC processes, only few publications are available on this field. In (Ludemann-Homburger et al., 2000), the VARICOL principle is introduced and a first approach to the description of the discrete port positioning is presented. The approach is used to determine the mean number of columns between the inlet and outlet ports. The concept is extended in (Hanisch, 2002; Toumi et al., 2002a,b) to generally describe the port switching of the VARICOL. Time–invariant switching patterns are considered. The dynamics of the port switching is represented in from of algebraic equations and lookup tables. A diagram representation of the mean column numbers between the ports is introduced. Until now, no model of the discrete dynamics considering time–varying switching patterns has been published.

The third topic concerns the combined discrete and continuous dynamics, which are governed by switchings in the continuous dynamics and jumps of the continuous states. Selected publications, which concern the modelling of this class of hybrid dynamical systems, are (Kamau, 2004; Lunze, 2002; v. d. Schaft and Schumacher, 2000). In (Kleinert and Lunze, 2002), the first approach to the distinct modelling of the continuous and discrete dynamics of the SMB is published. In this thesis, the continuous and the discrete dynamics are modelled separately. The discrete dynamics are represented by means of finite deterministic automata. The combined discrete and continuous dynamics is represented using the concept of hybrid automata (Alur et al., 1993). The hybrid nature with respect to the interaction of the discrete and the continuous dynamics is analysed.

State reconstruction of distributed parameter systems. If the controlled variables cannot be measured directly, a reconstruction from available plant outputs is necessary. In case of SCC processes, the distributed and the switching nature of the parameters has to be considered. Especially if the product purity values are not measured directly, a major part of the concentration profiles has to be reconstructed in order to determine these controlled variables. In (Mangold et al., 1994) a concept is proposed based on a true counterflow model which uses continuously measured concentrations. In (Kloppenburg and Gilles, 1999; Kloppenburg, 2000), an extended Kalman Bucy filter approach is proposed, which uses a lumped–parameter model of the true counterflow. The use of a periodically time–varying Kalman filter is proposed in (Erdem et al., 2004), which is applied to a lumped–parameter model of the SMB. The periodicity of the observer error is considered to enhance the state reconstruction. In (Zimmer et al., 1999) a concept is proposed for the SMB, which uses the continuous–time component concentration measurements in the product outlets to determine the parameters of the physical model. Thereby, the concentration profiles are reconstructed. A recent work proposes the reconstruction of the concentration profiles by an online process simulation in combination with a determination of the model parameters (Toumi, 2005).

These approaches make a direct use of the physical model and, therefore, yield complex observation algorithms. The concepts based on the true counterflow model do not take into account the hybrid nature of the process.

In (Kleinert and Lunze, 2003), an approach for the reconstruction of parts of the SMB concentration profiles is presented. The observer model is based on the concepts of travelling waves in chemical separation processes (Marquardt, 1988, 1990). An explicit functional description of the characteristic physical model solution is used, of which the unknown parameters are determined by a state observer. The approach is extended using a linear time–varying discrete–time Luenberger observer (Ludyk, 1981), leading to a simple and easy to implement observation algorithm (Kleinert and Lunze, 2004, 2005). The approach uses selected discrete–time concentration measurements of the concentration profiles. It is detailed in this thesis.

Process control. SCC processes are complex multivariable systems with combined continuous and discrete dynamics. Hence, an intuitive approach to the control is to use a detailed physical model to determine the trajectory of the control input variables. A rigorous physical model of SCC processes is used in (Toumi, 2005) as a basis for a nonlinear model predictive control algorithm to determine the evolution of control input variables. Because of the complexity of the model, the necessity of online–determination of the model parameters and the state reconstruction, the complexity of the control algorithm is considerably high. The approach applies to SMB and to VARICOL processes and was tested on a real plant.

In (Natarajan and Lee, 2000), a repetitive model predictive control approach for SMB processes is presented, which makes use of the cyclicly occurring error of the process model. (Erdem et al., 2004) use this approach with a periodic Kalman filter for the reconstruction of the concentration profles based on the Convection–Diffusion–Equation as a physical SMB model. Although a spatial discretisation and linear model reduction techniques are applied the resulting observation and control algorithms are considerably complex. The implementation in a control unit and experimental results are described in (Abel et al., 2005).

In (Kloppenburg and Gilles, 1999; Kloppenburg, 2000) an exact input–output linearisation based on the spatially discretised model of the true counterflow is applied for the design of a state feedback control for the SMB. The concept uses discrete–time concentration measurements of the concentration profile.

In (Klatt et al., 2000) and (Hanisch, 2002), a two–layer control concept is proposed. The lower layer realises the control of the wave front positions by decentralised linear internal model controllers. An outer control loop is added for the control of the product purity values. The upper layer is used to identify the local controller design models in form of discrete–time transfer functions by numerical simulations of the physical model.

The reference (Schramm et al., 2003) proposes an inferential control of selected points of the concentration profile (see Section 1.3). In (Schramm et al., 2001, 2003) an approach is presented, which uses an analytic relation between the internal fluid flow rates and the wave front propagation velocities. The wave front velocities are prescribed by two PI controllers.

Only one of the concepts found in literature apply to SMB as well as to VARICOL processes. All other approaches refer to the control of SMB processes only. Despite of the fact that several control concepts have been published, no industrial application was reported until now.

1.5 Structure of the thesis

Chapter 2 gives a detailed description of the characteristics of SCC processes as a basis for the solution of the control tasks. The physical plant setup and the concepts of process behaviour representation are described. In Section 2.3.3 and 2.4.3 the relation between the concentration profiles and the product purity values is described. Based on this analysis, an operation mode for

SCC processes is proposed as the basis for the control concepts.

Chapter 3 presents the detailed physical modelling of the chromatographic continuous–variable fluid dynamical processes, the discrete–variable port switching and the combined discrete and continuous dynamics of SCC processes. In Section 3.2, the continuous dynamics are modelled based on the fluid dynamical single–column model, which is the Equilibrium–Dispersive–Model. The model is derived and the principle is applied to the derivation of the fluid dynamical TMB model in Section 3.3. An analytical expression for the stationary TMB model solution is presented. The model of the continuous dynamics of SCC processes is derived in Section 3.4. The discrete dynamics of the port switching is modelled by means of discrete–event systems (Section 3.5). The state–transition relation, the output relation, the automaton graph and the number of possible discrete states are derived for several kinds of SCC processes. The modelling of the hybrid dynamics by means of hybrid automata is presented in Section 3.5.3.

Chapter 4 describes a new concept for the reconstruction of the wave fronts of the SCC concentration profiles. In Section 4.1, the observation problem is described. A reduced order wave front model is derived in Section 4.2. The derivation of the observation algorithm is presented in Section 4.3 and an evaluation of the observation error is performed in Section 4.4.

Chapter 5 presents the continuous control of SCC processes. In Section 5.1, the control task is reformulated in terms of the SCC process model presented in Chapter 3. The design of measurement– and observation–based continuous SCC controllers is described in Sections 5.2 and 5.3. The principle of the system analysis and the model identification is introduced in Section 5.4. The application of the control concepts to three example SCC processes is shown in Section 5.5.

Chapter 6 introduces the combined discrete and continuous control of VARICOL processes. The control task for this control problem and the way of solution are described in Section 6.1 and 6.2. The switching rules of the discrete controller and the design of the continuous controller are presented in Section 6.3. An application example and numerical simulation results of the closed–loop behaviour are shown in Section 6.4.

A summary and an outlook are given in Chapter 7. The table of symbols, the data of the example separation processes, special derivations and experimental data are given in the appendices A, B and C. The author's curriculum vitae is given at the end of the thesis.

Chapter 2

Simulated counterflow chromatographic separations

In this chapter, the concept of simulated counterflow chromatography and the adsorption as one of its basic physical principles is described. The single–column separation, the true counterflow, the Simulated Moving Bed and the VARICOL principle are described and the concepts of process representation are introduced. It is discussed, how plant operation affects the product purity. The toolbox for the numerical simulation of SCC processes and the example separations, which are referred to in this thesis, are described.

2.1 Chromatographic separation

2.1.1 Analytic and preparative chromatography

Chromatographic separations are based on the different retention intensities of the components of a mixture towards an adsorbent in presence of a fluid. The adsorption is one of the retention mechanisms. Chromatographic separations are characterised by a high separation quality.

The application of chromatographic separations can be categorised into two classes:

1. The first class refers to the **analytic separation**. The aim is the separation of a mixture for the qualitative or quantitative determination of the single components of the mixture which shall be separated. For these applications, the single– or batch–column chromatography is applied.

2. The second class considers **preparative separations** and **production scale separations**. Here, the batch–column separation or the more complex implementation of Simulated Counterflow Chromatographic (SCC) separations is used. In the class of the counterflow processes, the most important principles are the Simulated Moving Bed process (SMB) and the Variable Length Column (VARICOL) process.

The focus of this thesis is the control of Simulated Counterflow Chromatographic separations. The basis for the technical implementation of these separations is the single–column chromatography. All chromatographic separations which are referred to in the following are adsorption based separations.

2.1.2 Adsorption behaviour

Adsorption–based chromatography uses the different adsorption tendencies of the components of a mixture towards the surface of an adsorbent. The components are dissolved in a solvent, which surrounds the adsorbent. The adsorption then takes place due to van–der–Waals forces between the adsorbent surface and the components (Lübbert, 2004).

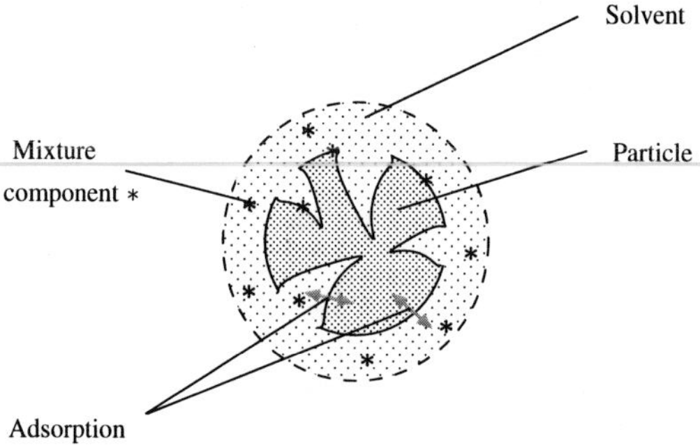

Figure 2.1: Mass exchange and adsorption between the solvent and the particle surface

Figure 2.1 shows a globular, porous particle of the adsorbent in the environment of the solvent. The solvent carries a dissolved component $*$. Due to the adsorption by the particle, a mass exchange of the component $*$ takes place between the solvent and the particle surface. The *adsorption equilibrium* is characterised by a stationary mass exchange. Then, the adsorbent is saturated with respect to the amount of the component which is locally available in the solvent.

Usually, several different adsorbable components $* \in \{A, B, \dots\}$ take part in the adsorptive mass exchange. The equilibrium relation between the local component concentrations c_* in the solvent and q_* on the adsorbent surface is, in the general case, determined by a function $f_{Eq,*}$ with

$$q_* = f_{Eq,*}(c_A, c_B, \dots) .$$

Because the adsorption behaviour depends upon the adsorption temperature, the parameters of $f_{Eq,*}$ are constant for a given temperature and $f_{Eq,*}$ is called the adsorption isotherm.

The adsorption behaviour of the separation processes, which are considered in this thesis, are assumed to be described by the equilibrium function

$$q_* = f_{Eq,*}(c_*) \,, \tag{2.1}$$

where q_* is only a function of c_*. In the simplest case, when low concentrations c_* can be considered, q_* is a linear function of c_* with

$$q_* = H_* \, c_* \,.$$

H_* is the linear adsorption coefficient of the component $*$. It is called the Henry–coefficient. For the general case, the adsorption is described by the so–called Hill isotherm which is given by

$$q_* = \frac{q_{s,*}}{n} \frac{b_{1,*} \, c_* + 2 \, b_{2,*} \, c_*^2 + \cdots + n \, b_{n,*} \, c_*^n}{1 + b_{1,*} \, c_* + b_{2,*} \, c_*^2 + \cdots + b_{n,*} \, c_*^n} \,. \tag{2.2}$$

The parameters $q_{s,*}$, $b_{1,*}$, $b_{2,*}$, ... , $b_{n,*}$ are unique for the component $*$ and for a given temperature, and n is the order of the isotherm. For $n = 3$, the cubic Hill isotherm is obtained:

$$q_* = \frac{q_{s,*}}{3} \frac{b_{1,*} \, c_* + 2 \, b_{2,*} \, c_*^2 + 3 \, b_{3,*} \, c_*^3}{1 + b_{1,*} \, c_* + b_{2,*} \, c_*^2 + b_{3,*} \, c_*^3} \,. \tag{2.3}$$

Because $f_{Eq,*}$ is a nonlinear function, the adsorption equilibrium described by $f_{Eq,*}$ is referred to as the nonlinear adsorption. Two general cases of nonlinear adsorption are distinguished:

1. the saturation of the adsorbent, which is referred to as the *Langmuir–like* adsorption, and
2. the increasing adsorption tendency, which is considered as the *Anti–Langmuir–like* adsorption.

In the first case, the adsorbent concentration q_* converges to a constant value for an increasing solvent concentration c_*. An example of a Langmuir–like isotherm is represented in the left plot of Figure 2.2 using the cubic Hill isotherm. In the second case, the slope of the isotherm increases with an increasing solvent concentration c_* due to an increasing adsorption tendency. The right plot of Figure 2.2 shows an example of an Anti–Langmuir–like isotherm. This example is also described by the cubic Hill isotherm.

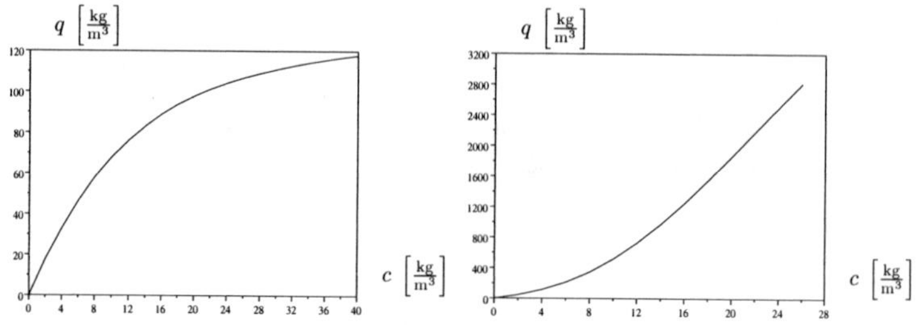

Figure 2.2: Langmuir–like and Anti–Langmuir–like Hill isotherms

2.1.3 Single–column separation

The simplest implementation of the chromatographic separation is the single–column separation. The principle uses a pipe, which is called the *separation column* and which is filled with a fixed bed of porous particles. The particles have an adsorptive surface and are called the *adsorbent*. All components of the mixture which is to be separated show different adsorption intensities towards the adsorbent surface.

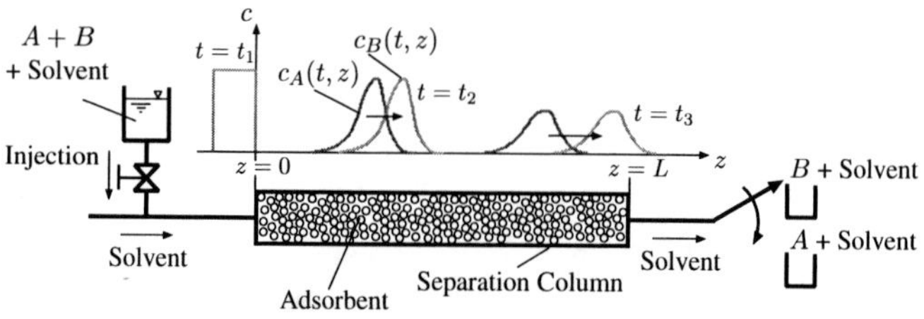

Figure 2.3: Chromatographic separation in a single separation column

For the separation of the mixture, a *solvent* is pumped through the porous bed of the adsorbent. The mixture, which is dissolved in the solvent, is injected discontinuously into the solvent stream at the column inlet (see Figure 2.3). The solvent carries the components, which shall be separated, through the adsorbent. When the components get in contact with the adsorbent surface, adsorption and desorption take place simultaneously. Thereby, the components propagate slower through the separation column than the solvent. Furthermore, because the adsorption intensities are different, the propagation velocities of the components are different. The separation of the components takes place because the distance between the components increases.

The single–column separation is a fluid dynamical process including different kinds of mass transport and mass exchange. The process behaviour is described by the spatial and temporal distribution of the components in the separation column. Figure 2.3 shows the separation of a mixture with two components A and B. Component A has the stronger, and component B has the weaker adsorption tendency towards the adsorbent.

The distribution of the components in the column are represented by the concentration profiles $c_A(t, z)$ and $c_B(t, z)$, where t is the global time and $z \in [0, L]$ is the spatial coordinate which is defined over the length L of the separation column. In the moment of the injection at time $t = t_1$, the concentration profiles have the form of pulses. Due to wall friction, remixing effects in the fixed bed and the adsorptive mass exchange, the pulses are broadened and degenerate to bell curves or waves. With increasing time t the profiles propagate through the separation column. Due to the different propagation velocities of the components, the distance between the waves increases (compare the concentration waves at $t = t_2$ and $t = t_3$). At the outlet, the solvent stream is directed into different containers to collect the separated components.

The representation of the solvent concentration values in the outlet stream of the separation column at $z = L$ over the time t is referred to as the *chromatogram*. Figure 2.4 shows an example, where the signal of an ultra violet (UV) light detector represents the concentration values. The chromatogram shows three waves with different maximum values. Each wave indicates the outlet concentration curve of one component.

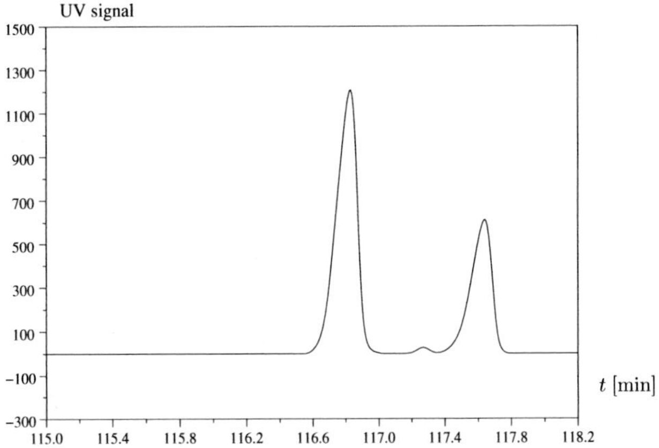

Figure 2.4: Example chromatogram of a three component mixture separation

When the separation is terminated, the containers contain the volume of V_A or V_B of solvent. In these volumes, each a certain amount of partial masses m_* of the component $*$ and $m_{b,*}$ of the respective impurity is dissolved ($b, *$ is used in the following to specify the by–product of the component $*$ in case of a binary preparative separation). The partial masses are determined by the volume V_* and the concentrations c_* and $c_{b,*}$:

$$\begin{aligned} m_* &= V_* \, c_* \\ m_{b,*} &= V_* \, c_{b,*} \, . \end{aligned} \tag{2.4}$$

In the preparative chromatographic separation the purity r is an important measure for the separation quality. The purity values for the previously discussed binary single–column separation are given by

$$\begin{aligned} r_A &= \tfrac{m_A}{m_A + m_{b,A}} \\ r_B &= \tfrac{m_B}{m_B + m_{b,B}} \, . \end{aligned} \tag{2.5}$$

Applying Equation (2.4) to Equation (2.5) yields the well known relation between the component concentrations and the corresponding purity values:

$$\begin{aligned} r_A &= \tfrac{c_A}{c_A + c_{b,A}} \\ r_B &= \tfrac{c_B}{c_B + c_{b,B}} \, . \end{aligned}$$

The single–column separation process allows to achieve very high purity values. The process is characterised by the discontinuous injection of the feed mixture and the discontinuous product recovery.

2.1.4 Continuous chromatography

To overcome the discontinuity of preparative single–column separations, the concept of continuous chromatographic separations was developed, which enables the continuous feed injection and the continuous recovery of the products. The concept is based on the counterflow of the adsorbent and the solvent. Because the true counterflow cannot be implemented it has to be simulated by a relative movement of the inlet and outlet ports with respect to the fixed adsorbent. This principle is called a Simulated Counterflow Chromatographic process, or briefly SCC process. Relevant industrial SCC processes are the Simulated Moving Bed and the Variable–Length–Column process. Both are implementations of the virtual True Moving Bed process, which constitutes the true counterflow. The following Sections 2.2, 2.3 and 2.4 describe the principle and the characteristics of these continuous chromatographic separation processes.

2.2 True Moving Bed process

2.2.1 Counterflow principle

The True Moving Bed (TMB) process is based on the true counterflow between fluid, i.e. the solvent, which carries the mixture component $*$, and the porous bed of adsorbent, in a separation column. Figure 2.5 shows an example where the fluid is fed into the column on the left side and flows in the opposite direction of the adsorbent. The fluid stream leaving the column is recycled, mixed with pure solvent and redirected to the column. The adsorbent is fed to the right inlet of the column and propagates in the opposite direction of the internal fluid flow. It is supposed that also the adsorbent is recycled and redirected to the column.

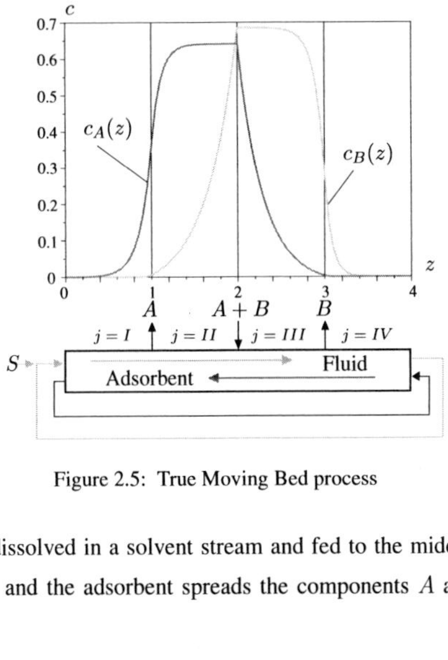

Figure 2.5: True Moving Bed process

The mixture $A + B$ is dissolved in a solvent stream and fed to the middle of the column. The counterflow of the fluid and the adsorbent spreads the components A and B over the column interior.

The component A is supposed to have the stronger adsorption tendency towards the adsorbent. Hence, it is mainly carried in the direction of the adsorbent propagation. The component B with the weaker adsorption tendency is mainly carried in the direction of the fluid flow. The two components propagate continuously into opposite directions. The product outlet ports are placed at a position where the concentration of *one* component is predominant. Through the outlet ports, a part of the internal fluid flow is withdrawn to recover the separated components of the mixture.

The column is divided into four sections, which are delimited by the column inlet, the outlet ports of A and B, the feed inlet port $A + B$, and the column outlet. The sections of the TMB are indicated by $j = I, II, III, IV$. The internal fluid flow rate is different in each section. To compensate the solvent loss which arises due to the product streams, pure solvent is fed to the recycled fluid via the solvent inlet port S.

The TMB process is virtual because the transport of the adsorbent can not be implemented in practice. The TMB concept is introduced to understand and to model the chromatographic counterflow principle. Furthermore, it is the basis for the implementation of the counterflow principle. For a better understanding of the simulated counterflow, the TMB process with a closed–loop separation column is introduced. It corresponds to the process setup shown in Figure 2.6, where the ends of the separation column are connected such that the column forms a loop, in which the fluid circulates in the opposite direction of the adsorbent.

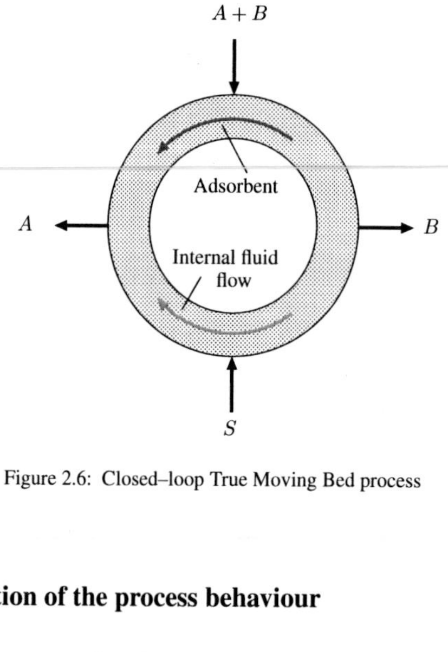

Figure 2.6: Closed–loop True Moving Bed process

2.2.2 Representation of the process behaviour

Like for the single–column separation, the components of the mixture are distributed in the separation column. The distributions are represented by the concentrations $c_A(t, z)$ and $c_B(t, z)$ of A and B in the fluid. Because of the real counterflow of the fluid and the adsorbent, the steady state concentration profiles $c_A(z)$ and $c_B(z)$ are time–invariant functions of the spatial coordinate z. Figure 2.5 shows an example, where $z \in [0, 4]$ is defined over the four sections of the TMB. The origin of the spatial coordinate system is placed at the position of the solvent inlet S. The concentrations c_A and c_B in each TMB section are described by the same class of functions, with different parameters for each section and component.

2.2.3 Product purity

In contrast to the single–column separation the TMB is a continuous process. Hence, the products are provided by the continuous fluid flows $\dot{m}_A$ and $\dot{m}_B$, which are withdrawn from the column at the product outlet ports A or B. Throughout this thesis, $\dot{m}_A$ denotes the extract mass flow is rare and $\dot{m}_B$ denotes the raffinate mass flow. Each flow consists of a solvent stream which carries different amounts of the components A and B. The extract flow carries more of the component A and the raffinate flow carries more of the component B.

In the stationary operation mode the product flow rates are constant. Because the component concentration values are time–invariant in the stationary state, the product concentrations c_* and the by–product concentrations $c_{b,*}$ in the product streams are also constant.

As for the single–column separation, the purity values are the important measures for the classification of the separation performance. The purity values are determined by the relation of the partial masses of the products m_* and the by–products $m_{b,*}$ at the outlet port of the product $* = A, B$ (Equation (2.5)). In the case of the continuous separation the partial masses have to be determined from the partial mass flows $\frac{\dot{m}_*}{\rho_s} c_*(z_*)$ of the product and $\frac{\dot{m}_*}{\rho_s} c_{b,*}(z_*)$ of the by–product. ρ_s is the solvent density and z_* is the outlet position of the product $*$. The partial masses are obtained by the integration of the partial mass flow over a given time horizon T:

$$\begin{aligned} m_* &= \int_0^T \frac{\dot{m}_*}{\rho_s} c_*(z_*)\, d\tau \\ m_{b,*} &= \int_0^T \frac{\dot{m}_*}{\rho_s} c_{*,b}(z_*)\, d\tau\,. \end{aligned} \tag{2.6}$$

Because $\dot{m}_*$, $c_*(z_*)$ and $c_{b,*}(z_*)$ are constant in the stationary state, the TMB product purity values in the steady state are determined by

$$\begin{aligned} r_A &= \frac{c_A(z_A)}{c_A(z_A)+c_{b,A}(z_A)} \\ r_B &= \frac{c_B(z_B)}{c_B(z_B)+c_{b,B}(z_B)}\,. \end{aligned}$$

2.3 Simulated Moving Bed process

2.3.1 Implementation of the counterflow principle

Simulated counterflow. The simulation of the counterflow between the internal fluid and the adsorbent is achieved by moving the inlet and outlet ports in the direction of the internal fluid flow, instead of moving the adsorbent in the opposite direction. With respect to the positions of the inlet and outlet ports, the same counterflow occurs as with the TMB process. This principle realises the continuous imitation of the true counterflow (Figure 2.7 (a)). The adsorbent is kept in a fixed porous bed and its movement is simulated by moving the inlet and outlet ports.

It is impossible to implement the continuous movement of the inlet and outlet ports. Therefore, the column is divided into several single separation columns, which are connected to a circle by capillaries or tubes. The ports can be attached to any of the column interconnections. Figure 2.7 (b) shows the setup, where the circle consists of eight columns in total. The process is configured such that there are each two columns between two ports. The port movement is realised by a stepwise and synchronous shifting of all inlet and outlet ports by one column length in the direction of the internal fluid flow. The principle is called Simulated Moving Bed (SMB). It was developed in the 1960s and was patented by UOP (Broughton and Gerhold, 1961). The time span $[0, T_S]$ between two port switchings is called the switching period or simply the period. T_S is called the switching time.

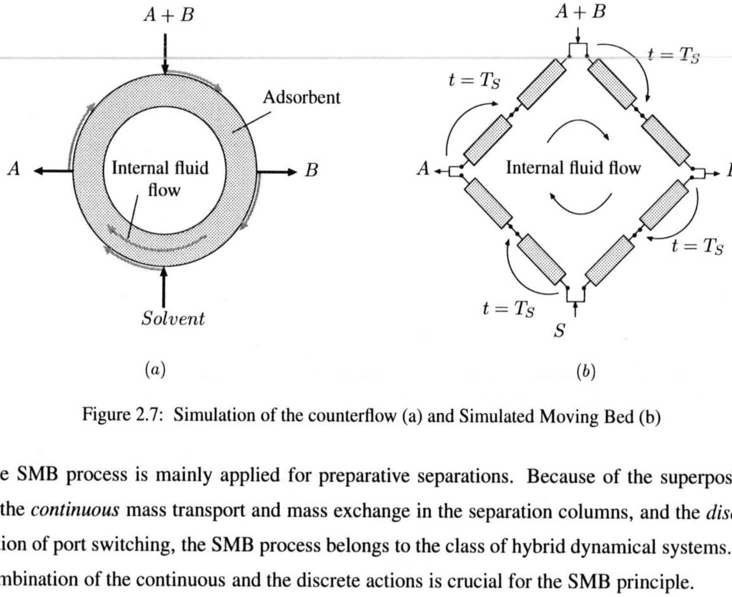

Figure 2.7: Simulation of the counterflow (a) and Simulated Moving Bed (b)

The SMB process is mainly applied for preparative separations. Because of the superposition of the *continuous* mass transport and mass exchange in the separation columns, and the *discrete* action of port switching, the SMB process belongs to the class of hybrid dynamical systems. The combination of the continuous and the discrete actions is crucial for the SMB principle.

Configurations of SMB processes. The SMB process is, like the TMB, divided into four spatial sections $j = I, II, III, IV$ which are delimited by the positions of the inlet and outlet ports. The configuration of an SMB process is determined by the number of columns in the sections. The process shown in Figure 2.7 (b) has two columns per SMB section. This configuration is described by the abbreviation (2/2/2/2). The total number of columns in an SMB process is n_c. In this example, $n_c = 8$. Common configurations besides of this one are (1/2/2/1) or (2/3/3/2). Because the adsorbent is usually very cost intensive the number of separation columns should be reduced as much as possible. For the SMB process, the minimum number of columns is four, using the (1/1/1/1) configuration. However, this case leads to rather low product purity values. Therefore, the most commonly used configurations are (1/2/2/1) or (2/2/2/2).

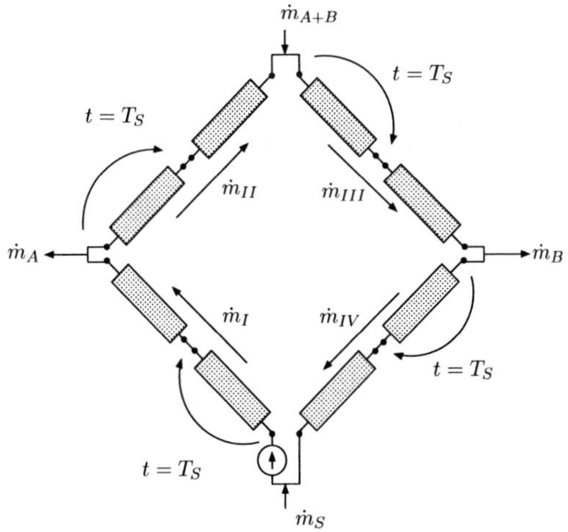

Figure 2.8: Operation parameters of the SMB process

Operation parameters. The operation parameters of an SMB process are the switching time T_S and the internal fluid flow rates $\dot{m}_I$, $\dot{m}_{II}$, $\dot{m}_{III}$ and $\dot{m}_{VI}$ in the SMB sections (see Figure 2.8). These parameters are control input candidates of the process. Because a fluid stream is injected or withdrawn at each inlet and outlet port, the internal fluid flow rates are different in each SMB section. The fluid streams at the inlet and outlet ports are $\dot{m}_S$, $\dot{m}_A$, $\dot{m}_{A+B}$ and $\dot{m}_B$, which are called the *external fluid flow rates*. If one of the internal fluid flow rates is given, all remaining internal fluid flow rates are determined by applying predetermined values to three of the external flow rates. Several possibilities exist for an implementation of this concept. Considering a pump mounted in section IV, which forces the internal fluid flow rate $\dot{m}_{IV}$ to a given value, the internal fluid flow rates in sections I, II and III are determined by the external fluid flow rates $\dot{m}_S$, $\dot{m}_A$ and $\dot{m}_{A+B}$:

$$\begin{aligned} \dot{m}_I &= \dot{m}_{IV} - \dot{m}_S \\ \dot{m}_{II} &= \dot{m}_I - \dot{m}_A \\ \dot{m}_{III} &= \dot{m}_{II} + \dot{m}_{A+B} \,. \end{aligned} \tag{2.7}$$

Equation (2.7) can be transformed to

$$\begin{aligned} \dot{m}_S &= \dot{m}_I - \dot{m}_V \\ \dot{m}_A &= \dot{m}_I - \dot{m}_{II} \\ \dot{m}_{A+B} &= \dot{m}_{III} - \dot{m}_{II} \,. \end{aligned} \tag{2.8}$$

Hence, in this setup the flow rates $\dot{m}_S$, $\dot{m}_A$, $\dot{m}_{A+B}$ have to be set according to Equation (2.8) in order to obtain the desired internal fluid flow rates $\dot{m}_I$, $\dot{m}_{II}$ and $\dot{m}_{III}$. SMB plants are constructed in a way that allows to adjust the internal fluid flow rates independently in a broad range.

With respect to the recovery of the recycling stream $\dot{m}_R = \dot{m}_{IV}$, which is the outlet stream of the SMB section IV, two different plant configurations are possible. The first configuration is referred to as the *closed recycling loop* where the recycling stream $\dot{m}_R$ is redirected to the inlet of the section I, as shown in Figure 2.7 or 2.8. The second configuration is referred to as the *open recycling loop*, where $\dot{m}_R$ is not fed back but recovered for further use in a separate container. Then, pure solvent is fed to the inlet of section I and the solvent inlet stream $\dot{m}_S = \dot{m}_I$ is equal to the internal fluid flow of the section I.

Disturbances. SCC processes underly disturbances which affect the adsorption behaviour, the internal fluid flow velocities or the feed inlet concentration values. In the following, some common disturbances are listed:

1. A temperature change affects the adsorption behaviour, which means a variation of the adsorption isotherm parameters.

2. The adsorbent underlies aging effects. Due to abrasive effects the adsorbing property of the adsorbent surface is reduced.

3. Varying pressure, which is caused by the shutdown of the unit and, to a moderate amount, by the port switching, reduces the porosity of the fixed bed while the mass of the adsorbent remains the same. The consequence is that the length of the fixed bed in the separation columns is reduced. Hence, the surface available for the adsorption of the components is reduced and the propagation velocity of the internal flow increases for constant internal fluid flow rates. Both effects, the reduced adsorption surface and the increasing flow velocity,

lead to a drift of the concentration profiles in the direction of the internal fluid flow and, hence, at least to a decrease of the purity of B.

4. The amount of injected feed can vary during process operation, e.g. by step changes of the feed inlet concentration. This leads to a transient of the SCC process which changes the product purity.

5. The column interconnections consist of tubes or capillaries, which constitute a volume between the separation columns where no adsorption takes place. These volumes cause time delays in the transmission of the mixture components from one column to the next and are referred to as the extra–column volumes. The extra–column volumes might have different values for each column interconnection e.g. due to an internal fluid flow pump mounted on one interconnection.

6. The fixed bed porosity of the adsorbent shows certain deviations for each separation column and it changes during process operation. Hence, the adsorption has different characteristics in each column.

Within this work, two disturbances of SCC processes are considered. On the one hand, it is assumed that a step disturbance of the feed inlet concentrations can occur. On the other hand, a temporal drift of the package porosity over a given time horizon is considered. All further effects, especially the different extra–column volumes of the column interconnections and the deviation of the package porosities of the separation columns, are assumed to be negligible.

2.3.2 Representation of the process behaviour

As for the single–column separation and the TMB process, the behaviour of the SMB process is described by the distribution of the two components A and B in the separation columns. The distributions are represented by the concentrations c_A and c_B, which describe the local amount of the dissolved components A and B in the internal fluid flow. In the stationary state of the SMB process, the concentration profiles show a similar shape compared to the TMB concentrations. However, because the adsorbent is brought into the columns in form of a fixed package, the component distributions are forced to propagate through the circle of separation columns by the internal fluid flow. Therefore, the concentration profiles are time–varying in the stationary state and are always a function of time t and space z:

$$
\begin{aligned}
c_A &= c_A(t, z) \\
c_B &= c_B(t, z)\,.
\end{aligned}
\tag{2.9}
$$

An intuitive representation of the concentration profiles is obtained by using a moving spatial coordinate system, where the origin $z = 0$ is placed at the position of the solvent inlet S. When the ports are switched, the coordinate system is moved with the ports by one column length such that the origin $z = 0$ remains at the position z_S of the solvent inlet. This concept is referred to as the *moving spatial coordinate*. For the representation of the concentration profiles c_A and c_B it is suitable to introduce the *local time counter*

$$\tau \in [0, T_S]$$

as an independent variable of the switching period. Using k as the counter of the switching periods, the global time t can be expressed in terms of τ and k by

$$t = \tau + k\,T_S$$

and k is also an independent variable. A switching cycle K (simply called a cycle) is completed after n_c switchings, where n_c is the number of columns of the SMB process. In other words, after each cycle K, all inlet and outlet ports have performed n_c switchings and, hence, arrive at their initial position on the circle of columns.

The third independent variable is the coordinate z of the moving spatial coordinate system. It is defined over the length L of the n_c separation columns. The column length L is normalised to $L = 1$. Hence, z is defined over the interval $z \in [0, n_c]$. With the variables τ, z and k, the representation (2.9) of the concentration profiles can be replaced by

$$\begin{aligned} c_A &= c_A(\tau, z, k) \\ c_B &= c_B(\tau, z, k)\,. \end{aligned} \tag{2.10}$$

Figure 2.9 shows an example of this representation. The dark gray curves show the concentration profile of the component A and the light gray curves show the profile of the component B. The figure shows snapshots of the profiles for different values of τ within one switching period. Shortly after a port switching instant at $\tau = 0$, the concentration profiles take a "left" position on the spatial coordinate system z. Due to the internal fluid flow, the profiles move along the spatial coordinate z with increasing τ, until the next port switching takes place at $\tau = T_S$. Then, because of the shifting of the spatial coordinate, the profiles are moved backwards in z by one column length $L = 1$.

Using the representation (2.10) of the concentration profiles based on the local time counter τ, the moving coordinate system z, and the counter of the switching periods k, the similarities between

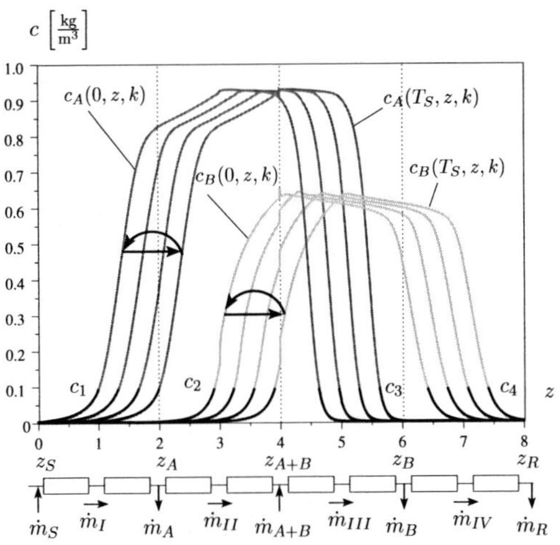

Figure 2.9: Concentration profiles in an SMB plant

the TMB and the SMB process become obvious. On the one hand, the positions of the inlet and outlet ports with respect to z are always the same for the TMB and the SMB. On the other hand, the characteristic bell shape (or wave) of the concentration profiles of the TMB is also found in the SMB.

Neglecting different column package porosities and extra–column volumes, the cyclic movement of the stationary profiles occurs with respect to one switching period k. Then, two snapshots of the concentration profiles, which are recorded at the local time point $\tau' \in [0, T_S]$ in two subsequent switching periods, have the same shape

$$\begin{aligned} c_A(\tau', z, k) &= c_A(\tau', z, k+1) \\ c_B(\tau', z, k) &= c_B(\tau', z, k+1) \end{aligned} \tag{2.11}$$

in the stationary operation mode. In the general case, the same stationary shape is obtained after a full switching cycle K:

$$\begin{aligned} c_A(\tau', z, k) &= c_A(\tau', z, k+n_c) \\ c_B(\tau', z, k) &= c_B(\tau', z, k+n_c)\,. \end{aligned}$$

Wave fronts in SMB processes. Within this work, the wave fronts are considered as those parts of the concentration profiles, which have low concentration values. In Figure 2.9 the wave fronts are the lower parts of the concentration profiles, which are marked by the black lines. The wave fronts c_1, c_2, c_3, and c_4 are an important part of the concentration profiles. For the derivation of several published observation and control concepts, the SMB behaviour is described using the concept of the wave fronts (Kienle, 1997; Klatt et al., 2002; Schramm et al., 2001; Zimmer et al., 1999). The importance of the wave fronts lies in the fact of delimiting the spatial expansion of the concentration profiles with respect to z: The wave fronts c_1 and c_3 determine the spatial expansion of c_A, and the wave fronts c_2 and c_4 determine the spatial expansion of c_B. Like the concentration profiles, the wave fronts are represented in terms of τ, z and k:

$$\begin{aligned} c_1 &= c_1(\tau, z, k) \\ c_2 &= c_2(\tau, z, k) \\ c_3 &= c_3(\tau, z, k) \\ c_4 &= c_4(\tau, z, k)\,. \end{aligned}$$

The wave fronts are categorised into desorption and adsorption wave fronts. In the separation columns, through which the wave fronts c_1 and c_2 propagate, the desorption of the components A and B by the internal fluid streams $\dot{m}_I$ and $\dot{m}_{II}$ takes place. Hence, the wave fronts c_1 and c_2 are *desorption wave fronts*. In the separation columns, through which the wave fronts c_3 and c_4 propagate, an adsorption of the components takes place. Hence, c_3 and c_4 are called the *adsorption wave fronts*.

2.3.3 Product purity

A product purity of 100 % is not always reasonable. Some applications require a precise product purity below 100 %. Furthermore, a purity value below 100 % allows for a more economical operation of the plant. Nevertheless, a lower purity limit has to be guaranteed to obtain a prescribed product quality. Two aspects have to be considered with respect to the purity value:

1. The minimum purity requirements limit the performance of the separation with respect to the feed throughput (Storti et al., 1993; Kloppenburg and Gilles, 1998; Schramm et al., 2002a).

2. In case of disturbances, like a decreasing adsorbent package porosity or an increasing feed concentration, the purity values can be reduced significantly.

Considering these aspects it becomes clear that an open–loop control of SMB processes cannot guarantee a prescribed product quality. Hence, it is necessary to analyse how the concentration profile propagation and the port switching influences the product purity. Furthermore, it is necessary to analyse how the concentration profiles and thereby the purity values can be manipulated by a variation of the operation parameters of the SMB process. In this section, these aspects are discussed in detail. It is presented, how the purity values are determined from the concentration profiles and how the purity values can be manipulated by changing the propagation of the wave fronts.

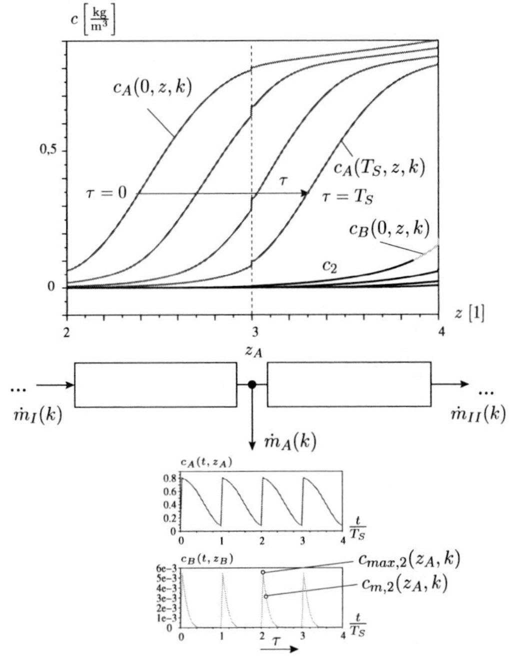

Figure 2.10: Evolution of the product and by–product concentrations at the outlet of A during one switching period with $\tau \in [0, T_S]$

Impurities in the product stream. The wave fronts are of special interest for the design and the control of SCC separations because they determine the impurities in the product streams $\dot{m}_A$ and $\dot{m}_B$ and the recycling stream $\dot{m}_R$. Considering the outlet of A at z_A in Figure 2.9 shows that if, due to a low internal flow rate $\dot{m}_{II}$, the wave front c_2 does not move far enough along the z-axis during one switching period k, the wave front c_2 steps back too far in the direction of the outlet of A in the moment of port switching. Then, the impurity of the product stream $\dot{m}_A$ by the component B increases (clearly, in a real plant, the concentration profiles do not step backwards in the moment of port switching; this effect occurs due to the modelling concept of the moving spatial coordinate z).

Figure 2.10 shows a close–up of the concentration profiles at the outlet position z_A. It shows, how the concentrations $c_A(\tau, z, k)$, $c_B(\tau, z, k)$ and the wave front $c_2(\tau, z, k)$ evolve during one switching period. It is possible to specify the impurity of $\dot{m}_A$ by one concentration value $c_{m,2}(z_A, k) = c_2(\tau_m, z_A, k)$, which is recorded at the measurement time $\tau_m \in [0, T_S]$ at the position z_A. If $\tau_m = 0$, $c_{m,2}$ represents the largest value $c_{max,2}(z_A, k)$ of c_2, which occurs at z_A. It is the same concentration that occurs at $\tau = T_S$ at the position $z_A + 1$ before the port switching:

$$c_{max,2}(z_A, k) = c_2(0, z_A, k) = c_2(T_S, z_A + 1, k - 1)\,. \tag{2.12}$$

The diagrams in the lower part of Figure 2.10 show the repetitive temporal evolution of c_A and c_B at the position z_A. The diagrams show the recorded measurement $c_{m,2}(z_A, k)$ and the value $c_{max,2}(z_A, k)$ in the evolution of $c_2(t, z_A)$.

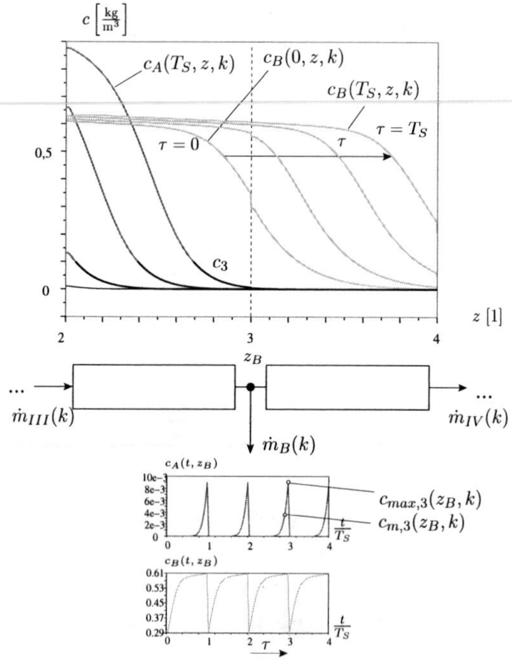

Figure 2.11: Evolution of the product and by–product concentrations at the outlet of B during one switching period with $\tau \in [0, T_S]$

With respect to a suitable operation mode the following rule can be formulated:

Limiting $c_{m,2}$ to a certain value by adjusting the internal flow $\dot{m}_{II}$ means limiting the impurity of the product A.

On the other hand, it is also possible to vary the switching instant to avoid the impurity of $\dot{m}_A$ by c_2. This, however, refers to the more elaborate concept of asynchronous port switching, which is described in Section 2.4.

The product stream $\dot{m}_B$ is recovered at the position z_B. Figure 2.9 shows how the impurity of $\dot{m}_B$ at the position z_B is influenced by the wave front c_3. Figure 2.11 shows a close–up of the concentration profiles at the position z_B. If the internal fluid flow rate $\dot{m}_{III}$ is too large such that the wave front c_3 propagates too far in the direction of z_B, the impurity of the product stream $\dot{m}_B$ by the component A increases. The degree of the impurity is indicated by the concentration value $c_{m,3}(z_B, k) = c_3(\tau_m, z_B, k)$, $\tau_m \in [0, T_S]$ of the wave front c_3 at the position z_B. For $\tau_m = T_S$, $c_{m,3}(z_B, k)$ is the largest concentration of c_3, which occurs at the position z_B. Because of the repeated movement of the concentration profiles along the SMB sections, the same concentration occurs at $\tau = 0$ at the position $z_B - 1$ in the next switching period $k + 1$:

$$c_{max,3}(z_B, k) = c_3(T_S, z_B, k) = c_3(0, z_B - 1, k + 1)\,. \tag{2.13}$$

This means that for process operation the following rule can be applied:

Limiting $c_{m,3}$ by the corresponding choice of $\dot{m}_{III}$ means limiting the impurity of $\dot{m}_B$.

Recycling stream impurity. The wave fronts c_1 and c_4 also contribute to the impurity of the product streams. This is shown in Figure 2.12. It shows the same concentration profiles as Figure 2.9, however, with the solvent inlet at a centred position. If the internal fluid flow rate $\dot{m}_{IV}$ is too large, the wave front c_4 propagates too far in the direction of z_S and spreads out into the section I. Because $\dot{m}_I$ is larger than $\dot{m}_{IV}$, the wave front moves on through section I until it reaches the outlet of A at z_A and contributes to the impurity of the product stream $\dot{m}_A$. This means that a partial amount of B, which reaches the position z_S and pours into the recycling stream $\dot{m}_R$, is carried through section I to the outlet of A. The degree of this impurity is indicated by the concentration value $c_{m,4}(z_S, k) = c_4(\tau_m, z_S, k)$, $\tau_m \in [0, T_S]$ at the position z_S, which refers to the impurity of the recycling stream $\dot{m}_R$ by c_4. If $\tau_m = T_S$, $c_{m,4}(z_S, k)$ is the largest concentration of c_4 which occurs at the position z_S. At the position $z_S - 1$, the same concentration occurs at $\tau_m = 0$ in the next switching period $k + 1$:

$$c_{max,4}(z_S, k) = c_4(T_S, z_S, k) = c_4(0, z_S - 1, k + 1)\,. \tag{2.14}$$

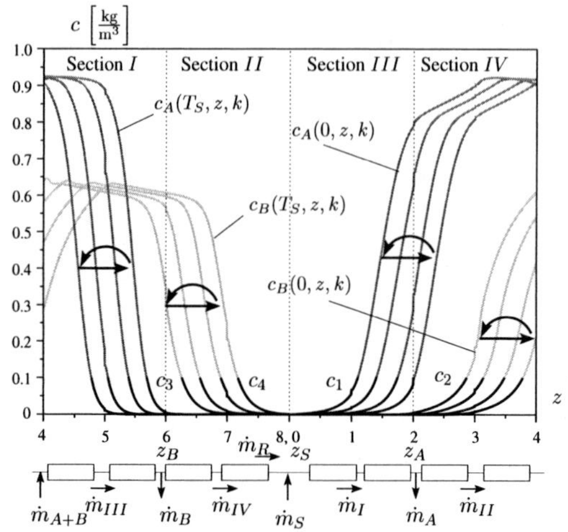

Figure 2.12: Evolution of the concentration profiles and wave fronts with respect to z_S

The wave front c_1 contributes to the impurity of the product stream $\dot{m}_B$, if the internal fluid flow rate $\dot{m}_I$ is too low. Then, the wave front c_1 does not propagate far enough into the SMB section I during one switching period and is switched back in the moment of port switching. Thus, the wave front partly reaches out into the section IV. The internal fluid flow rate $\dot{m}_{IV}$ dilutes the wave front from the SMB section IV through the recycling stream into section I. However, because $\dot{m}_{IV}$ is smaller than $\dot{m}_I$, the wave front propagation of c_1 in section IV is even slower than in section I such that with each port switching, the wave front steps further in the direction of z_B where it contributes to the impurity of the product stream $\dot{m}_B$. This means that the larger the impurity of $\dot{m}_R$ by c_1 is, the larger the contribution of c_1 to the impurity of $\dot{m}_B$ becomes. The degree of the impurity is indicated by the concentration value $c_{m,1}(z_S, k) = c_1(\tau_m, z_S, k)$, which is recorded at $\tau_m \in [0, T_S]$ at the position z_S. If $\tau_m = 0$, then $c_{m,1}(z_S, k)$ is the largest concentration of c_1, which occurs at the position z_S. The same value $c_{max,1}(z_S, k)$ also occurs at $\tau = T_S$ at the position $z_S + 1$ before the port switching:

$$c_{max,1}(z_S, k) = c_1(0, z_S, k) = c_1(T_S, z_S + 1, k - 1)\,. \qquad (2.15)$$

Product purity values. The important variables of the SMB process, which determine the quality of the separation, are the purity values r_A and r_B at the product outlets. The purity values are determined by the partial masses m_A and m_B of the components A and B, which are recovered at the outlets z_A and z_B of the products:

$$\begin{aligned} r_A &= \left.\frac{m_A}{m_A+m_B}\right|_{z=z_A} \\ r_B &= \left.\frac{m_B}{m_A+m_B}\right|_{z=z_B} . \end{aligned} \tag{2.16}$$

The purity computation for the single–column separation considers a constant solvent volume V. The volume contains a constant dissolved amount of the components A and B, which is represented by the component concentrations c_A and c_B. The partial masses are given by

$$\begin{aligned} m_A &= c_A V \\ m_B &= c_B V . \end{aligned}$$

In the present case of a continuous separation, the partial masses must be determined from the product streams $\dot{m}_*$, $* = A, B$. Because the dynamical transition of the SMB process is of interest, time-varying product streams, which depend upon τ and k, are considered first to describe the general case:

$$\begin{aligned} \dot{m}_A &= \dot{m}_A(\tau, k) \\ \dot{m}_B &= \dot{m}_B(\tau, k) . \end{aligned}$$

The component concentrations c_A and c_B at the product outlet z_* are time–varying:

$$\begin{aligned} c_A(z_*) &= c_A(\tau, z_*, k) \\ c_B(z_*) &= c_B(\tau, z_*, k) . \end{aligned}$$

The partial masses are determined by the integration of the partial mass flow over a given time horizon. A suitable time horizon considers one switching period k with $\tau \in [0, T_S]$. Hence, the following expression for the partial masses is obtained, where ρ_S is the solvent density:

$$\begin{aligned} m_A(z_*, k) &= \frac{1}{\rho_S} \int_0^{T_S} \dot{m}_*(\tau, k)\, c_A(\tau, z_*, k)\, d\tau \\ m_B(z_*, k) &= \frac{1}{\rho_S} \int_0^{T_S} \dot{m}_*(\tau, k)\, c_B(\tau, z_*, k)\, d\tau . \end{aligned}$$

If it is considered that the product stream $\dot{m}_* = \dot{m}_*(k)$ is constant during one switching period, the partial masses are given by

$$\begin{aligned} m_A(z_*, k) &= \frac{1}{\rho_S} \dot{m}_*(k)\, \bar{c}_A(z_*, k) \\ m_B(z_*, k) &= \frac{1}{\rho_S} \dot{m}_*(k)\, \bar{c}_B(z_*, k) , \end{aligned} \tag{2.17}$$

where $\bar{c}_A(z_*, k)$ and $\bar{c}_B(z_*, k)$ are the mean concentrations of the components A and B in the period k at the position z_*:

$$\begin{aligned} \bar{c}_A(z_*, k) &= \frac{1}{T_S} \int_0^{T_S} c_A(\tau, z_*, k)\, d\tau \\ \bar{c}_B(z_*, k) &= \frac{1}{T_S} \int_0^{T_S} c_B(\tau, z_*, k)\, d\tau\,. \end{aligned} \tag{2.18}$$

The application of Equation (2.17) to Equation (2.16) shows how the purity values r_A and r_B of the SMB process are determined by the mean concentrations:

$$\begin{aligned} r_A(k) &= \frac{\bar{c}_A(z_A, k)}{\bar{c}_A(z_A, k) + \bar{c}_B(z_A, k)} \\ r_B(k) &= \frac{\bar{c}_B(z_B, k)}{\bar{c}_A(z_B, k) + \bar{c}_B(z_B, k)}\,. \end{aligned} \tag{2.19}$$

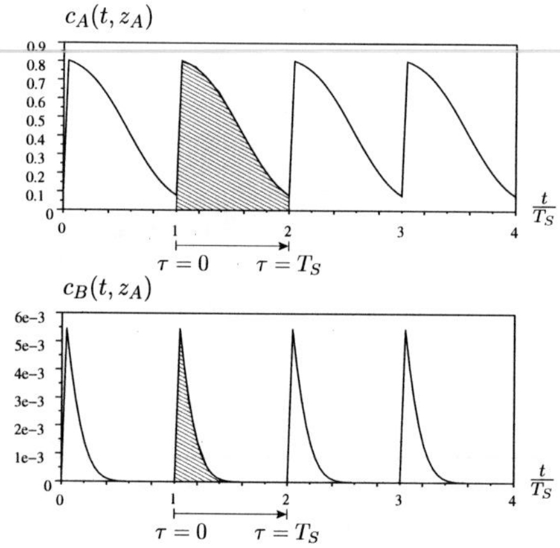

Figure 2.13: Temporal evolution of $c_A(\tau, z_A, k)$ and $c_B(\tau, z_A, k)$ in the stationary state

Product and by–product concentrations. Each product stream carries the product and also a certain amount the of by–product. With $* = A, B$, the resulting purity value of the product stream is represented using Equation (2.19):

$$r_*(k) = \frac{\bar{c}_*(z_*, k)}{\bar{c}_*(z_*, k) + \bar{c}_{b,*}(z_*, k)}\,. \tag{2.20}$$

$\bar{c}_*(z_*, k)$ specifies the mean product concentration, and $\bar{c}_{b,*}(z_*, k)$ specifies the mean by–product concentration at the position z_*. The component B is the by–product of the component A at $z = z_A$. Hence, $c_B(\tau, z_A, k)$ is the by–product concentration of the product A. In contrast, A is the by–product of B at the position $z = z_B$ and $c_A(\tau, z_B, k)$ is the by–product concentration of the product B at the position $z = z_B$. Considering the wave fronts of the SMB concentration profiles shows that the by–product at the outlet of A is determined by the wave fronts c_2 and c_4. The by–product at the outlet of B is determined by the wave fronts c_1 and c_3.

If the wave front c_1 does not reach out into the section IV and the wave front c_4 does not reach out in the section I, each of the by–products is primarily determined by *one* wave front: The by–product of A at the position z_A is mainly determined by $c_B(\tau, z_A, k) = c_2(\tau, z_A, k)$. At the position z_B, the by–product is mainly determined by $c_A(\tau, z_B, k) = c_3(\tau, z_B, k)$. To determine the purity $r_*(k)$ it is necessary to know the temporal evolution of the product *and* by–product concentrations at the position z_* (see Equations (2.18) and (2.19)). The temporal evolution of the concentrations in $\dot{m}_A$ is shown in Figure 2.13 for the stationary operation mode. The gray areas under the curves are a measure of the mean concentration values $\bar{c}_A(z_A, k)$ and $\bar{c}_B(z_A, k)$.

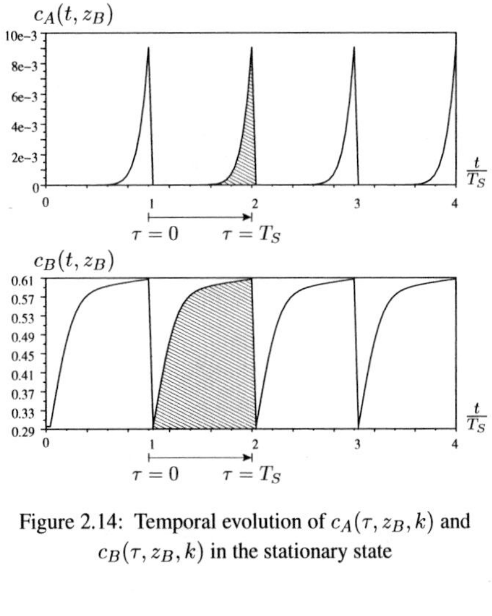

Figure 2.14: Temporal evolution of $c_A(\tau, z_B, k)$ and $c_B(\tau, z_B, k)$ in the stationary state

The temporal evolution of the concentrations in $\dot{m}_B$ is shown in Figure 2.14 for the stationary operation mode. The gray areas under the curves are a measure of the mean concentration values $\bar{c}_A(z_B, k)$ and $\bar{c}_B(z_B, k)$.

Operation mode for SMB processes. To conclude the discussion on the wave fronts c_i, $i = 1, 2, 3, 4$ and the purity values r_*, $* = A, B$ the following strategy is proposed as a suitable operation mode for SMB processes:

The wave front concentrations $c_{m,1}(z_S, k)$ and $c_{m,4}(z_S, k)$ are kept so low that the wave fronts c_1 and c_4 do not reach out into the sections IV or I, respectively. Then, the purity values r_A and r_B primarily depend upon the wave fronts c_2 and c_3.

Considering this operation mode it is easily possible to manipulate each wave front individually by the internal solvent mass flow rate of the corresponding SMB section, through which the wave front propagates. This fact is the justification for the implementation of a decentralised controller for the SMB as e.g. described in (Klatt et al., 2002; Hanisch, 2002; Schramm et al., 2003).

2.4 VARICOL process

In case of a separation, for which an expensive adsorbent has to be used, it is of interest to reduce the number of separation columns. A reasonable minimum of columns is $n_c = 4$. However, if an SMB process is implemented with four columns and a configuration (1/1/1/1), the purity values which can be achieved are lower than with a plant with at least two columns in the sections II and III, i.e. with a configuration (1/2/2/1). However, in a plant with four separation columns it is possible to obtain a mean of more than one column in the middle sections II and III and a mean of less than one column in the outer sections I and IV by applying the Variable Length Column (VARICOL) principle. This principle is based on the same concept as the SMB, with the difference that the inlet and outlet ports are switched *asynchronously*. It is commonly applied to plants with 4, 5 or 6 columns. A VARICOL process shows many similarities to the SMB process, however, it is necessary to introduce the concept of subperiods of the switching period k to describe the asynchronous switching of the ports. Furthermore, the dynamics of the discrete port switching shows a considerably higher complexity than that of the SMB.

The VARICOL principle was invented by NOVASEP in 2000 (Ludemann-Homburger et al., 2000). In recent publications, the performance and optimisation of the VARICOL is investigated (Toumi et al., 2002a,b). Until now, no concept for a closed–loop control of VARICOL processes has been published.

This section gives an introduction to the VARICOL principle and presents the characteristics of the process behaviour.

2.4.1 VARICOL principle

The VARICOL process also consists of a circle of separation columns, to which the inlet and outlet ports of the feed and the solvent inlet and the product outlets are connected. Like the SMB, the VARICOL process simulates the counterflow between the adsorbent and the internal fluid flow by a discrete shift of the inlet and outlet ports in the direction of the internal fluid flow. However, while all ports are switched synchronously on an SMB process, on a VARICOL process the ports are switched asynchronously. Figure 2.15 shows an example. The local time of the switching period of the solvent inlet S is represented by $\tau_S \in [0, T_S]$, where T_S is the duration of this switching period of the port S. The switching of S takes place at the end of this switching period at $\tau_S = T_S$. The switchings of the ports A, $A + B$ and B take place at the respective local times $\tau_S = \Delta T_A$, $\tau_S = \Delta T_{A+B}$ and $\tau_S = \Delta T_B$. These times are referred to as the relative switching times because they determine the switching instant of the ports A, $A + B$ and B relatively to the beginning of the switching period (see e.g. Figure 2.16). In the general case, the relative switching times ΔT_A, ΔT_{A+B} and ΔT_B are not equal.

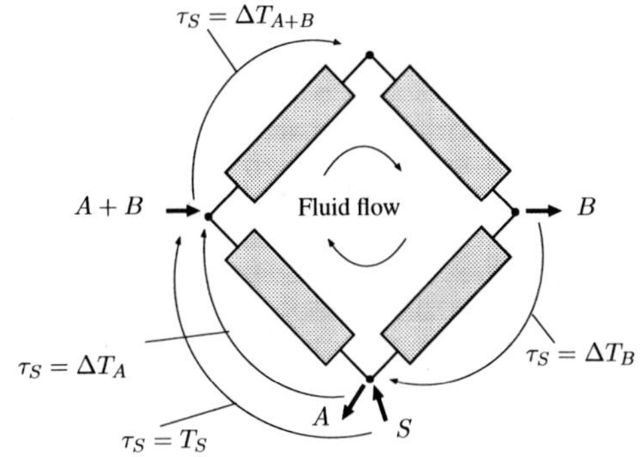

Figure 2.15: Asynchronous port switching of the VARICOL process

Switching pattern. The switching of the ports follows a given *pattern*. In principle it is possible to switch all ports independent of each other to any column interconnection of the circle of separation columns at any time. However, for the continuous separation of a binary mixture $A + B$ by the VARICOL principle, it is sufficient to consider a switching pattern, for which the following assumptions hold:

Assumption 2.4.1 *Assumptions for the switching patterns of VARICOL processes:*

1. *Starting with the position S and moving in the direction of the internal fluid flow, the inlet and outlet ports are arranged in the following order:*

$$S, A, A+B, B\,.$$

2. *All ports are successively switched in the direction of the fluid flow.*

3. *The ports cannot overtake each other, i.e. the order S, A, $A+B$, B is not changed.*

4. *If a port is switched, it is only switched by one column length.*

5. *There is always at least one column in the section III, i.e. between the ports $A+B$ and B.*

□

The following textbox summarises the meaning of the switching pattern:

A switching pattern of the VARICOL process is determined by the initial port position on the circle of separation columns given by the initial column configuration

$$(n_{c,0,I}/n_{c,0,II}/n_{c,0,III}/n_{c,0,IV})\,, \tag{2.21}$$

and the switching time T_S and the relative switching times ΔT_1, ΔT_2 and ΔT_3, with

$$0 \leq \Delta T_1 \leq \Delta T_2 \leq \Delta T_3 < T_S\,, \tag{2.22}$$

and their assignment to the ports A, $A+B$ and B.

For the formal description of the port switching, the switching of the solvent inlet S at the position z_S, which is switched at times $t = k_S\,T_S$, is used as a reference. k_S is the counter of the switching instants of the solvent inlet port. The assignment of ΔT_1, ΔT_2 and ΔT_3 to the switching of one of the ports A, $A+B$ and B determines the relative switching times ΔT_A, ΔT_{A+B} and ΔT_B.

Figure 2.16 shows an example. In this example, the inlet of $A+B$ is switched after a time period of $\Delta T_1 = \Delta T_{A+B}$ after the last switching of the solvent inlet S. The hereafter switched port is the outlet of A, and the last port is the outlet of B. This example shows, how the switching sequence of the ports is determined by the assignment of ΔT_1, ΔT_2 and ΔT_3 to the ports A, $A+B$ and B. To conclude, a switching pattern is determined by the initial configuration (2.21) of the VARICOL plant, the switching times (2.22) and their assignment to the ports A, $A+B$ and B. Based on the switching pattern, the mean section length can be determined, which is shown in the next after the following paragraph.

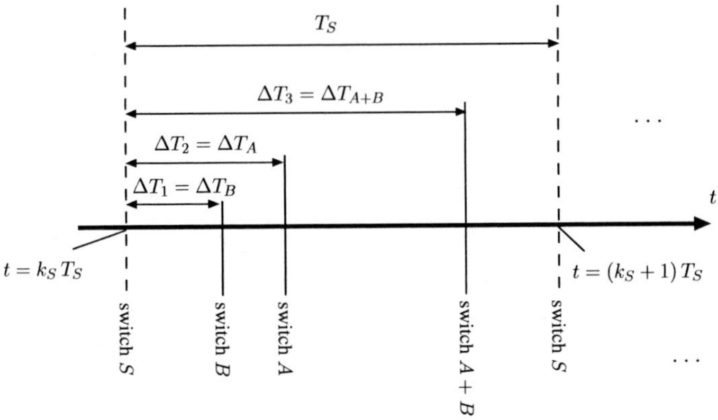

Figure 2.16: Switching times of the VARICOL process

Periods and sub–periods. Generally, it is possible to vary ΔT_1, ΔT_2, ΔT_3 and T_S during process operation. Furthermore, it is possible to vary the assignment of the relative switching times to the ports A, $A + B$ and B. Considering these degrees of freedom, there are, for a plant with $n_c = 4$ separation columns, 80 possible combinations of the port positions with respect to the separation columns, for which Assumption 2.4.1 holds (see Section 3.5.3). If, however, the assignment of ΔT_1, ΔT_2, ΔT_3 to the ports A, $A + B$ and B does not change, only $4 \cdot n_c$ combinations of port positions on the circle of separation columns occur. This means that in this case, only four possible column configurations $(n_{c,I}/n_{c,II}/n_{c,III}/n_{c,IV})$ occur independently of n_c. During one switching period T_S of the port S, these four configurations occur successively according to the assignment of the relative switching times to the ports. If, on the one hand, the assignment of the relative switching times ΔT_1, ΔT_2, ΔT_3 to the ports A, $A + B$ and B does not change and, on the other hand, ΔT_1, ΔT_2, ΔT_3 are constant in time, the switching pattern is said to be *time–invariant*. Because during one period of S four port switchings occur (including the switching of the port S), the period is subdivided into four sub–periods κ, which are labelled by $\kappa = 1, 2, 3, 4$, each with a unique column configuration. Hence, the VARICOL column configuration is a function of κ:

$$(n_{c,I}(\kappa)/n_{c,II}(\kappa)/n_{c,III}(\kappa)/n_{c,IV}(\kappa))\,.$$

The durations δT_κ of the sub–periods are determined by

$$\begin{aligned}
\delta T_1 &= \Delta T_1 \\
\delta T_2 &= \Delta T_2 - \Delta T_1 \\
\delta T_3 &= \Delta T_3 - \Delta T_2 \\
\delta T_4 &= T_S - \Delta T_3 \,.
\end{aligned}$$

Mean section length. For an SMB process, the number of columns $n_{c,j}$ in the section j is always a constant integer value. The number determines the configuration of the SMB plant with respect to the "length" of the SMB sections. For a VARICOL process, the (average) numbers of columns $\bar{n}_{c,j}$ in the single sections usually have non–integer values. $\bar{n}_{c,j}$ in the section j is determined by the column number $n_{c,j}(\kappa)$ and the durations δT_κ of the sub–periods of k_S (Hanisch, 2002; Toumi, 2005):

$$\bar{n}_{c,j}(k_S) = \sum_{\kappa=\kappa_0}^{\kappa_0+3} \frac{n_{c,j}(\kappa)\, \delta T_\kappa}{T_S(k_S)} \,. \tag{2.23}$$

$\bar{n}_{c,I}$, $\bar{n}_{c,II}$, $\bar{n}_{c,III}$ and $\bar{n}_{c,IV}$ are constant in case of a time–invariant switching pattern. The sum of all mean section lengths is equal to the total number of columns:

$$\bar{n}_{c,I} + \bar{n}_{c,II} + \bar{n}_{c,III} + \bar{n}_{c,IV} = n_c \,. \tag{2.24}$$

In the following, the mean section length for each section j is determined for a given switching pattern. The derivation is based on Equation (2.23) and is given in the Appendix C.3. $n_{c,j}(\kappa)$ and δT_κ are determined by the initial column configuration

$$(n_{c,0,I}/n_{c,0,II}/n_{c,0,III}/n_{c,0,IV})$$

the switching times T_S, ΔT_A, ΔT_{A+B} and ΔT_B of the considered switching period k_S. Hence, the mean section length $\bar{n}_{c,j}(k_S)$ is determined by the switching pattern of the considered VARICOL process:

$$\bar{n}_{c,I} = n_{c,0,I} + \frac{T_S - \Delta T_A}{T_S} \tag{2.25}$$

$$\bar{n}_{c,II} = n_{c,0,II} + \frac{\Delta T_A - \Delta T_{A+B}}{T_S} \tag{2.26}$$

$$\bar{n}_{c,III} = n_{c,0,III} + \frac{\Delta T_{A+B} - \Delta T_B}{T_S} \tag{2.27}$$

$$\bar{n}_{c,IV} = n_{c,0,IV} + \frac{\Delta T_B - T_S}{T_S}\,. \tag{2.28}$$

Equations (2.25) through (2.28) show, how the mean section lengths depend upon the switching pattern, i.e. on the switching time T_S, the relative switching times ΔT_A, ΔT_{A+B} and ΔT_B, and the initial column configuration $(n_{c,0,I}/n_{c,0,II}/n_{c,0,III}/n_{c,0,IV})$.

Based on the equations, the mean section length for the example shown in Figure 2.15 and 2.16 can be determined. The initial column configuration is (0/2/1/1). If the switching times shown in Figure 2.16 are applied with the values

$$\begin{aligned} T_S &= 125\text{ s} \\ \Delta T_B &= 40\text{ s} \\ \Delta T_A &= 85\text{ s} \\ \Delta T_{A+B} &= 115\text{ s}\,, \end{aligned}$$

the following mean section length configuration is obtained:

$$(0{,}32/1{,}76/1{,}6/0{,}32)\,.$$

Equations 2.25 through 2.28 can be transformed to determine the relative switching times: If T_S, the initial configuration $(n_{c,0,I}/n_{c,0,II}/n_{c,0,III}/n_{c,0,IV})$ and $\bar{n}_{c,I}$, $\bar{n}_{c,II}$ and $\bar{n}_{c,III}$ are given for a VARICOL with n_c columns, the relative switching times are determined by

$$\begin{aligned} \Delta T_A &= T_S\,(1 - \bar{n}_{c,I} + n_{c,0,I}) \\ \Delta T_{A+B} &= T_S\,(1 - (\bar{n}_{c,I} + \bar{n}_{c,II}) + (n_{c,0,I} + n_{c,0,II})) \\ \Delta T_B &= T_S\,(1 + \bar{n}_{c,IV} - n_{c,0,IV})\,. \end{aligned} \tag{2.29}$$

2.4.2 Representation of the process behaviour

The behaviour of the VARICOL process is, like the behaviour of the SMB, characterised by the distribution of the components in the circle of separation columns. This distribution is represented by the concentrations $c_A(t,z)$ and $c_B(t,z)$. Like for the SMB, it is convenient to use a spatial coordinate $z \in [0, n_c \cdot L]$, which origin $z_S = 0$ is tied to the position of the solvent inlet.

According to the concept of the switching period for the VARICOL, the concentration profiles are represented in terms of the local time counter τ_S, the spatial coordinate z and the switching period k_S:

$$\begin{aligned} c_A &= c_A(\tau_S, z, k_S) \\ c_B &= c_B(\tau_S, z, k_S)\,. \end{aligned}$$

The component distribution propagation in a 4–column VARICOL is represented in Figure 2.17 using four snapshots of the concentration profiles at different local time points τ_S. Because not only the concentration profiles, but also the inlet and outlet ports move along the spatial coordinate z during one switching period k_S, the four profile snapshots are recorded at the local times $\tau_S = 0$, $\tau_S = \Delta T_{A+B}$, $\tau_S = \Delta T_A$ and $\tau_S = \Delta T_B$ directly after the switching of the corresponding port. Each snapshot is represented in a separate plot, which shows the profiles and the port positions at the beginning of the corresponding sub–period κ. The composition of the four plots shows the superposition of the asynchronous port switching and the propagation of the concentration profiles during one switching period. The port positions, which are assigned to the corresponding sub–period κ, are indicated by the vertical lines.

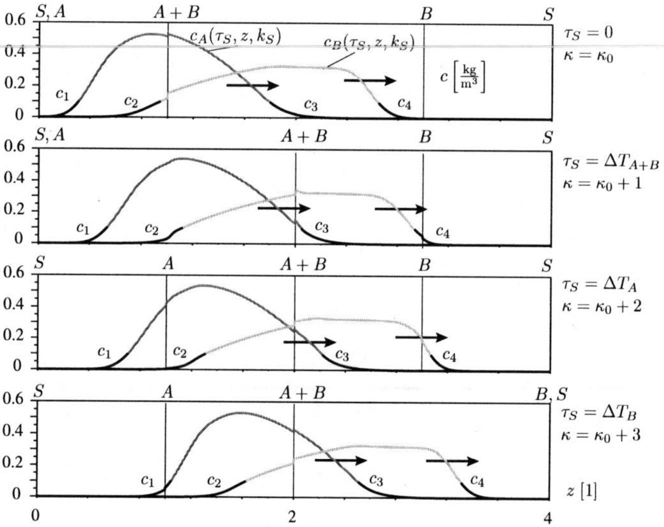

Figure 2.17: Stationary concentration profile snapshots during one switching period of the solvent inlet S

The spatial coordinate system is tied to the inlet of S and is shifted by one column length at the end of the switching period k_S, i.e. at $\tau_S = T_S$. Hence, at $\tau_S = T_S$ the concentration profiles perform a one column length jump to the left. Regarding the snapshot representation of Figure 2.17, this corresponds to a jump from the lower plot to the upper plot.

The stationary operation mode of the VARICOL process is determined by constant internal fluid flow rates and a time–invariant switching pattern. If the difference in the column package and the extra–column volumes between the separation columns are neglected, the movement of the steady state concentration profiles is repeated in every period. Then, the snapshots of the concentration profiles at a local time $\tau'_S \in [0, T_S]$ have the same shape for all subsequent periods $k_S, k_S + 1$:

$$\begin{aligned} c_A(\tau'_S, z, k_S) &= c_A(\tau'_S, z, k_S + 1) \\ c_B(\tau'_S, z, k_S) &= c_B(\tau'_S, z, k_S + 1) \,. \end{aligned}$$

If the difference in the column package porosity or the extra–column volumes is significant, the shape and movement of the concentration profiles show slight differences in two subsequent switching periods of the stationary operation mode. However, because the VARICOL is a periodic process, the movement and shape of the concentration profiles are the same in two subsequent switching *cycles*. The switching cycle of the VARICOL is defined in the same way as for the SMB.

Wave fronts in VARICOL processes. Like for the SMB process, the wave fronts c_1, c_2, c_3 and c_4 delimitate the VARICOL concentration profiles with respect to z. The wave fronts have the same meaning with respect to the product and the recycling stream impurity like it was discussed for the SMB in Section 2.3.3. Because the wave fronts are part of the concentration profiles, it is convenient to use the same representation (see Figure 2.17). Hence, the shape and the propagation of the wave fronts in one switching period k_S is described by

$$\begin{aligned} c_1 &= c_1(\tau_S, z, k_S) \\ c_2 &= c_2(\tau_S, z, k_S) \\ c_3 &= c_3(\tau_S, z, k_S) \\ c_4 &= c_4(\tau_S, z, k_S) \,. \end{aligned}$$

Remark. The representation of the concentration profiles and the wave fronts is not restricted to using the position of the solvent inlet as a reference. It is possible to use any of the ports S, A, $A + B$ or B as a reference and tie the spatial coordinate to this port. Then, the temporal evolution of the concentration profiles is represented with respect to that port, and the pair (τ_*, k_*), $* = S, A, A+B, B$ is used as independent variable, where k_* is the switching period and τ_* is the local time of k_* of the considered port $*$. The local time τ_* is defined over the interval $\tau_* \in [0, T_*]$, where T_* is the duration of the switching period k_*.

Process properties and operation parameters. The VARICOL process is clearly a process which encounters continuous and discrete dynamics. It shows the typical phenomena of hybrid dynamical systems: Using a moving spatial coordinate system representation, the concentration profiles perform a jump when the switching of the reference port takes place. Furthermore, with each switching of the ports the fluid flow velocity along the z-axis, which is characteristic for the continuous dynamics, is changed instantaneously in at least one column. Hence, state jumps and switching dynamics occur in the open–loop controlled plant.

The operation parameters of the VARICOL process are divided into two sets: the operation parameters P_c which determine the mass transport, and the operation parameters P_d which determine the switching of the ports. The first set consists of the internal fluid flow rates

$$P_c = \{\dot{m}_I, \dot{m}_{II}, \dot{m}_{III}, \dot{m}_{IV}\} ,$$

while the second set is given by the switching pattern

$$P_d = \{T_S, \Delta T_{A+B}, \Delta T_A, \Delta T_B, n_{c,0,I}, n_{c,0,II}, n_{c,0,III}, n_{c,0,IV}\} .$$

The parameters of P_c are continuous–valued. The parameters of P_d are continuous– or discrete–valued. The set P_c affects only the continuous part of the process, namely the mass transport and the mass exchange, whereas the set P_d determines the behaviour of the discrete part of the process, which is the port switching.

2.4.3 Product purity

The purity values of the VARICOL product streams are determined in the same way like the SMB purity values, namely by the product concentrations c_* and the by–product concentrations $c_{b,*}$ at the considered product outlet position z_*. The by–product concentrations represent the impurity of the product and are determined by the corresponding wave front concentrations. The following paragraphs show how the VARICOL purity values are determined by the wave fronts and show the analogy to the SMB.

Impurity of the product streams. Figure 2.18 shows a close–up view of the concentration profiles of the example process represented in Figure 2.17 in the area of the product outlet A. The spatial coordinate z_A is used as the reference here and, hence, the concentration profiles are represented in dependence upon the pair (τ_A, k_A). The figure shows how the concentrations

$c_A(\tau_A, z, k_A)$ and $c_B(\tau_A, z, k_A)$ propagate along the z–axis in the direction of the internal fluid flow. It shows the temporal evolution of the two concentrations at the postion z_A.

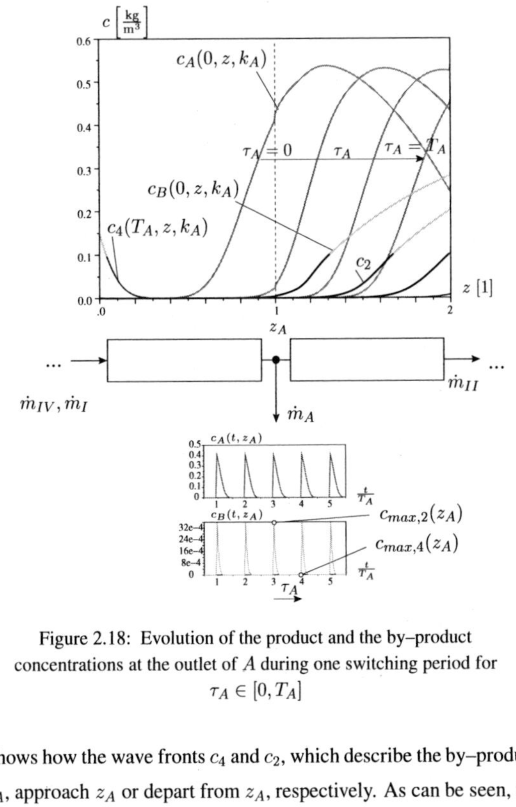

Figure 2.18: Evolution of the product and the by–product concentrations at the outlet of A during one switching period for $\tau_A \in [0, T_A]$

The figure also shows how the wave fronts c_4 and c_2, which describe the by–product concentration at the position z_A, approach z_A or depart from z_A, respectively. As can be seen, the wave front c_4 reaches out into the separation column, which is located at the upstream position of z_A: Shortly before the next switching of A the wave front takes the position $c_4(T_A, z, k_A)$. At this instant, c_4 causes the largest impurity of A, denoted by

$$c_{max,4}(z_A, k_A) = c_4(T_A, z_A, k_A)\,.$$

At $\tau_A = 0$, the wave front c_2 has a position for which the largest wave front concentration $c_{max,2}$ occurs at z_A:

$$c_{max,2}(z_A, k_A) = c_2(0, z_A, k_A)\,.$$

It can be seen from Figure 2.18 that $c_{max,2}$ is much larger than $c_{max,4}$, which means that the impurity of $\dot{m}_A$ is mainly determined by the wave front c_2 in this example.

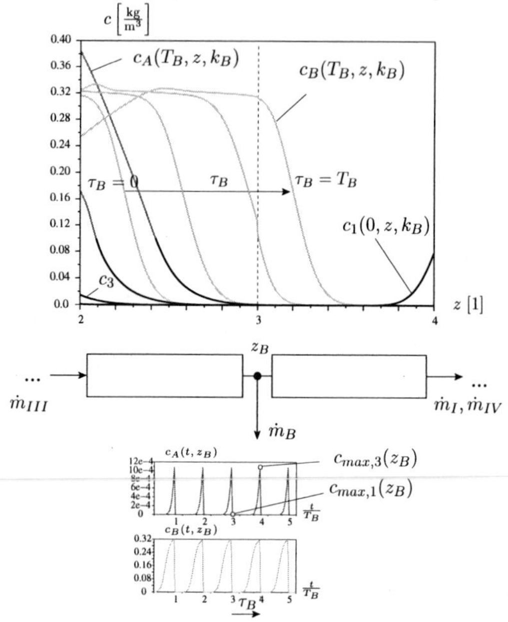

Figure 2.19: Evolution of the product and by–product concentrations at the outlet of B during one switching period $\tau_B \in [0, T_B]$

Figure 2.19 shows, how the concentrations c_A and c_B evolve in the area of the outlet of B. z_B and thereby the pair (τ_B, k_B) are chosen as the reference for the representation of the concentration profiles. The figure shows, how the concentration profiles move from their initial position at $\tau_B = 0$ to the position at $\tau = T_B$. The two plots in the lower part of the figure show the temporal evolution of c_A and c_B at the position z_B.

Figure 2.19 also shows, how the by–product concentration, which is determined by the wave fronts c_3 and c_1, contributes to the impurity of the product B. As can be seen, the wave front c_1 takes the initial position at $\tau_B = 0$ and then departs from the product outlet B. At the time point $\tau_B = 0$, the largest impurity of B by the wave front c_1 occurs. It is denoted by

$$c_{max,1}(z_B, k_B) = c_1(0, z_B, k_B)\,.$$

The figure also shows, how the wave front c_3 contributes to the impurity of B. At the time point

$\tau_B = 0$, the wave front takes the position with the largest distance to z_B. With increasing τ_B it approaches z_B and contributes to the impurity of B. The largest impurity by c_3 occurs at $\tau_B = T_B$. Then, the wave front c_3 takes its largest concentration value at the position z_B:

$$c_{max,3}(z_B, k_B) = c_3(T_B, z_B, k_B)\,.$$

Figure 2.19 shows that $c_{max,3}$ is much larger than $c_{max,1}$. Therefore, the impurity of $\dot{m}_B$ is mainly determined by the wave front c_3 in this example.

Impurity of the recycling stream. As can be seen in the Figures 2.18 and 2.19, the contribution of the wave fronts c_1 and c_4 to the impurity of the product streams is low compared to the contribution of the wave fronts c_2 and c_3. This is because the flow rates $\dot{m}_I$ and $\dot{m}_{IV}$ in the VARICOL sections I and IV are chosen such that c_1 does not reach out too far in the direction of z_B and c_4 does not propagate too far in the direction of z_A. However, it has to be considered that in a VARICOL process, the wave fronts c_1 and c_4 partly propagate through the sections I or IV, if sub–periods κ have to be considered for which $n_{c,I}(\kappa) = 0$ or $n_{c,IV}(\kappa) = 0$ occurs. During these sub–periods, the propagation of the respective wave front is determined by the fluid flow rates $\dot{m}_{II}$ or $\dot{m}_{III}$.

The concentrations of the wave fronts c_1 and c_4 contribute to the impurity of the recycling stream. Figure 2.20 shows the propagation of the concentration profiles and the wave fronts in the area of z_S during one switching period k_S. The concentration measurement

$$c_{m,4}(z_S, k_S) = c_4(\tau_m, z_S, k_S)\,, \quad \text{for } \tau_m \in [0, T_S]$$

can be used to value the impurity of the recycling which is caused by c_4. If $\tau_m = T_S$, then $c_{m,4}$ is the largest concentration value of the wave front c_4 at the position z_S:

$$c_{max,4}(z_S, k_S) = c_4(T_S, z_S, k_S)\,.$$

The larger the value of $c_{m,4}(z_S, k_S)$ is, the higher becomes the impurity of the recycling, and thereby of the product A, by the wave front c_4.

The concentration measurement

$$c_{m,1}(z_S, k_S) = c_1(\tau_m, z_S, k_S)\,, \quad \text{for } \tau_m \in [0, T_S]$$

indicates the impurity of the recycling by c_1. If $\tau_m = 0$, $c_{m,1}(z_S, k_S)$ is the largest concentration of c_1 at the position z_S:

$$c_{max,1}(z_S, k_S) = c_1(0, z_S, k_S)\,.$$

A large concentration $c_{m,1}(z_S, k_S)$ leads to a high impurity of the recycling and thereby of the product stream of B by the wave front c_1.

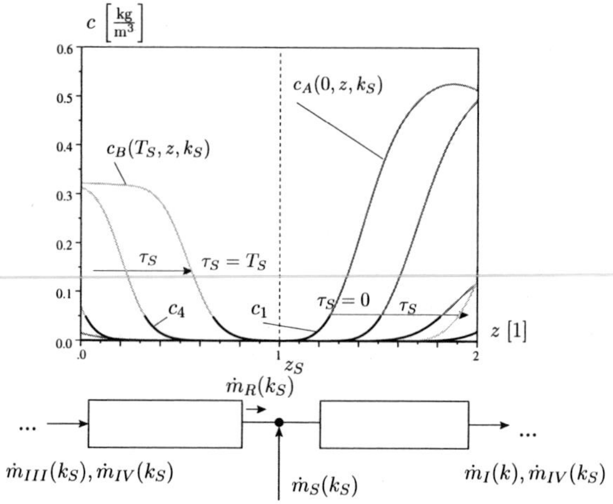

Figure 2.20: Evolution of the VARICOL concentration profiles and wave fronts with respect to z_S

Product purity values. The product purity values are, as for the SMB process, determined by the partial masses of the product and the by–product in the corresponding product streams. To determine the purity values, a temporal balance horizon over one switching period k_S is applied. Using Equation (2.16) with

$$\begin{aligned} r_A &= \left.\frac{m_A}{m_A+m_B}\right|_{z=z_A} \\ r_B &= \left.\frac{m_B}{m_A+m_B}\right|_{z=z_B}, \end{aligned} \tag{2.30}$$

the partial masses m_A and m_B are determined by

$$
\begin{aligned}
m_A(z_*, k_S) &= \tfrac{1}{\rho_S} \int\limits_0^{T_S} \dot{m}_*(\tau_S, k_S)\, c_A(\tau_S, z_*, k_S)\, d\tau_S \\
m_B(z_*, k_S) &= \tfrac{1}{\rho_S} \int\limits_0^{T_S} \dot{m}_*(\tau_S, k_S)\, c_B(\tau_S, z_*, k_S)\, d\tau_S \,,
\end{aligned}
$$

where T_S is the duration of the switching period k_S. z_* is the position of the considered outlet port $* = A, B$. If the internal fluid flow rates are changed e.g. by a controller only in the moment of port switching at $\tau = T_S$, the product stream $\dot{m}_*(\tau_S, k_S)$ is constant during the switching period k_S and the partial masses are given by

$$
\begin{aligned}
m_A(z_*, k_S) &= \tfrac{T_S}{\rho_S}\, \dot{m}_*(k_S)\, \bar{c}_A(z_*, k_S) \\
m_B(z_*, k_S) &= \tfrac{T_S}{\rho_S}\, \dot{m}_*(k_S)\, \bar{c}_B(z_*, k_S) \,,
\end{aligned}
$$

where $\bar{c}_A(z_*, k_S)$ and $\bar{c}_B(z_*, k_S)$ are the mean concentrations of the components A and B over the period k_S at the position z_*:

$$
\begin{aligned}
\bar{c}_A(z_*, k_S) &= \tfrac{1}{T_S} \int\limits_0^{T_S} c_A(\tau, z_*, k_S)\, d\tau \\
\bar{c}_B(z_*, k_S) &= \tfrac{1}{T_S} \int\limits_0^{T_S} c_B(\tau, z_*, k_S)\, d\tau \,.
\end{aligned}
$$

Applying these equations to Equation (2.30) shows how the purity values r_A and r_B are determined by the mean concentrations:

$$
\begin{aligned}
r_A(k_A) &= \tfrac{\bar{c}_A(z_A, k_A)}{\bar{c}_A(z_A, k_A) + \bar{c}_B(z_A, k_A)} \\
r_B(k_B) &= \tfrac{\bar{c}_B(z_B, k_B)}{\bar{c}_A(z_B, k_B) + \bar{c}_B(z_B, k_B)} \,.
\end{aligned}
\tag{2.31}
$$

Using $* = A, B$, the resulting purity value of the product stream $\dot{m}_*$ is derived using Equation (2.31):

$$
r_*(k_S) = \frac{\bar{c}_*(z_*, k_S)}{\bar{c}_*(z_*, k_S) + \bar{c}_{b,*}(z_*, k_S)} \,, \tag{2.32}
$$

with

$$
\begin{aligned}
\bar{c}_*(z_*, k_S) &= \tfrac{1}{T_S} \int\limits_0^{T_S} c_*(\tau_S, z_*, k_S)\, d\tau_S \\
\bar{c}_{b,*}(z_*, k_S) &= \tfrac{1}{T_S} \int\limits_0^{T_S} c_{b,*}(\tau_S, z_*, k_S)\, d\tau_S \,.
\end{aligned}
\tag{2.33}
$$

Equations (2.31), (2.32) and (2.33) show the analogy to the purity determination of the SMB process described in Section 2.3.3.

Operation mode for VARICOL processes. A suitable operation mode for VARICOL processes is now proposed based on the analysis of the wave front impact on the product purity values. The operation mode corresponds to that of SMB processes proposed in Section 2.3.3:

> The wave front concentrations $c_{m,1}(z_S, k_S)$ and $c_{m,4}(z_S, k_S)$ are kept low such that the contribution of the wave fronts c_1 and c_4 to the impurity of the the product streams $\dot{m}_A$ and $\dot{m}_B$ is negligible. Then the impurity of the product streams is primarily determined by the wave fronts c_2 and c_3.

Considering this operation mode it is possible to manipulate each of the wave fronts by the variation of the internal fluid flow rate of the corresponding VARICOL section, through which the wave front mainly propagates. Furthermore, the purity values can easily be adjusted by the suitable manipulation of $\dot{m}_{II}$ and $\dot{m}_{III}$.

2.5 Concentration and purity measurement

Concentration measurements. The state of the separation unit is determined from measurements of the internal concentration profiles.

For SMB and VARICOL plants the measurement of the internal concentration profile is only possible between the separation columns. Two measurement configurations can be applied:

1. On the one hand, discrete–time concentration measurements in the column interconnections can be considered. This technique uses the analytical single–column separation (Depta et al., 1999). The time point τ_m of the measurement can be chosen from the interval $\tau_m \in [0, T_*]$, where T_* is the duration of the switching period of the considered port $* = S, A, A+B, B$. This measurement principle provides discrete concentration values of high accuracy. However, it has a low sampling rate, e.g. one measurement per switching period. Figure 2.21 shows an example of the measurement of the wave fronts c_1 and c_2. At the measurement position $z_{m,1} = 1$ the wave front concentration $c_{m,1}(z_{m,1}, k_*) = c_1(\tau_{m,1}, z_{m,1}, k_*)$ is recorded at the measurement time $\tau_{m,1}$ in each switching period k_* of the respective port $*$. Correspondingly, the wave front concentration $c_{m,2}(z_{m,2}, k_*) = c_2(\tau_{m,2}, z_{m,2}, k_*)$ is recorded at the position $z_{m,2} = 3$. The measurement positions are fixed with respect to the

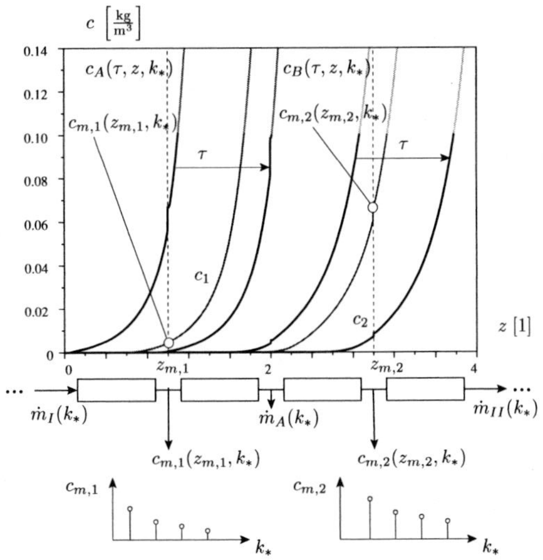

Figure 2.21: Examples of a discrete–time concentration measurements on an SMB or VARICOL plant

sections and, therefore, are moved with each port switching of the respective port $*$ by one column length.

2. On the other hand, continuous–time concentration measurements in the column interconnections of the plant are possible (see e.g. (Hanisch, 2002; Erdem et al., 2004; Toumi, 2005)). For this technique, a spectroscopic detector is mounted in the column interconnection and records the single component concentrations or the sum of the concentration values (depending on the detector technology). Figure 2.22 shows an example, where the wave front concentrations $c_1(\tau_*, z_{m,1}, k_*)$ and $c_2(\tau_*, z_{m,2}, k_*)$, $\tau_* \in [0, T_*]$ are continuously measured at the positions $z_{m,1}$ and $z_{m,2}$. For the example in Figure 2.22, the measurement positions are moved with the coordinate system in the moment of port switching. Besides of mounting a continuous measurement in the column interconnections it is also possible to install such a unit in the product outlet stream (Abel et al., 2005).

Both the discrete and the continuous concentration measurement principles apply to the SMB as well as to the VARICOL. The concentration measurements are crucial for the control concepts introduced in this thesis. The discrete–time concentration measurement is used for the continuous control of SCC processes (Chapter 5). The continuous–time concentration measurement is applied for the discrete control of the VARICOL port switching (Chapter 6).

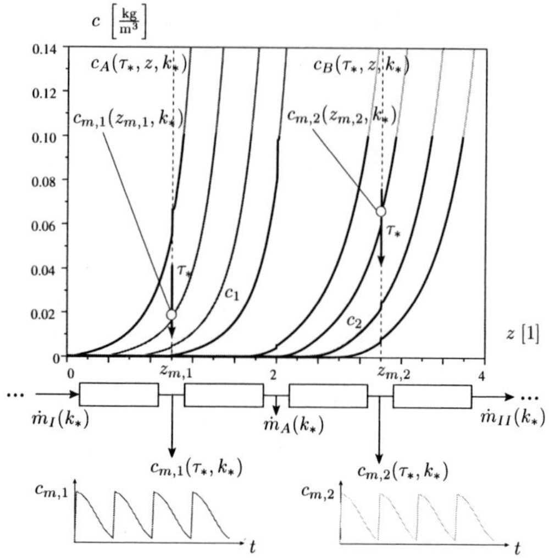

Figure 2.22: Examples of a continuous concentration measurements on an SMB or VARICOL plant

Purity measurement. As shown in the previous section, the purity values of SMB and VARICOL processes are determined by the mean concentration values of the product $\bar{c}_*(z_*, k_S)$ and the by–product $c_{b,*}(z_*, k_S)$ at the respective product outlet port z_*. If during one switching period the product stream $\dot{m}_*(k_S)$ is collected in a container, the solvent mass contained in the container is

$$m(z_*, k_S) = T_S\, \dot{m}_*(k_S)\,, \quad \text{for } * = A, B\,.$$

If the content is ideally mixed, the component concentrations in the container are the mean concentrations $\bar{c}_*(z_*, k_S)$ and $c_{b,*}(z_*, k_S)$. Hence, a single analytic concentration measurement at the end of each switching period in each product outlet is sufficient to determine the product purity $r_*(k_S)$, $* = A, B$. Figure 2.23 shows a scheme of the measurement setup.

Because the purity measurement has to be performed at the end of the switching period k_S, the measurement result $r_*(k_S)$ is available during the subsequent period $k_S + 1$, depending on the setup of the mixture analysis unit. With respect to the control of SCC processes this means that a time–delay of one switching period has to be considered for the purity measurement.

The concentration and the purity measurement concepts apply to the SMB and the VARICOL. For the SMB, k_* or k_S has to be replaced by k.

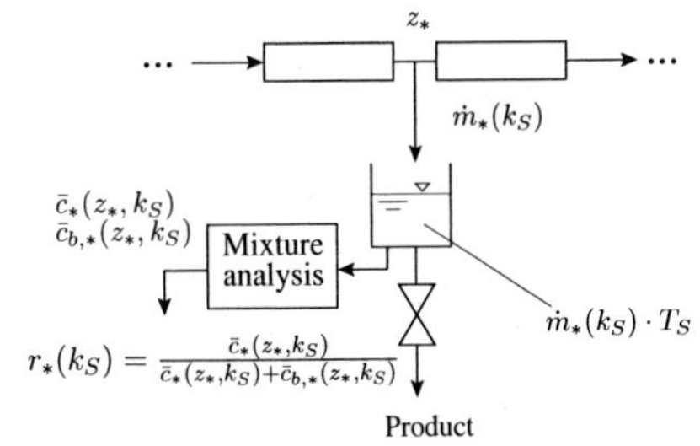

Figure 2.23: Discrete product purity measurement

2.6 Numerical simulation of chromatographic processes

Because only parts of the internal concentration profiles of SMB and VARICOL processes are available by measurements and because experiments on these processes are time consuming and difficult to perform, a numerical simulation of SMB and VARICOL processes based on physical models was used to analyse the stationary and dynamical behaviour. To provide a means for the simulation of single–column, SMB and VARICOL processes for various plant configurations, measurement configurations and separation problems, the CSep Toolbox for MATLAB was developed (Kleinert, 2002). CSep stands for Chromatographic Separation. To assure the correctness of the simulation results, sample chromatograms of the considered plant columns were used for the identification of the physical model parameters.

The CSep Toolbox provides finite–difference solvers for the Equilibrium–Dispersive–Model and the Equilibrium–Dispersive–Transport–Model for compressible and incompressible solvents (see e.g. (Giese, 2002; Kleinert, 2002)). Three kinds of adsorption isotherms are implemented: the linear isotherm, the cubic Hill isotherm and the Langmuir isotherm for the description of the concurrent adsorption of two components. Further features of the simulation are:

- ◇ SMB and VARICOL processes with four up to theoretically an infinite number of separation columns,
- ◇ separations with several components,
- ◇ variable and different extra–column volumes of the SCC column interconnections,
- ◇ variable and different package porosities in the SCC separation columns as well as variable and different column parameters and
- ◇ open or closed recycling loops for the SMB or the VARICOL.

The basic concept of the toolbox was developed by the author during his PhD work at the Institute of Control Engineering at Technical University Hamburg–Harburg. The main extensions and the main part of the implementation, which makes up the current version of the toolbox, was performed by the author during the continuation of the PhD at the Institute of Automation and Computer Control at the Ruhr–Universität Bochum. All plots of concentration evolutions and profiles shown in this thesis, except of the experimental data, are generated using the CSep toolbox.

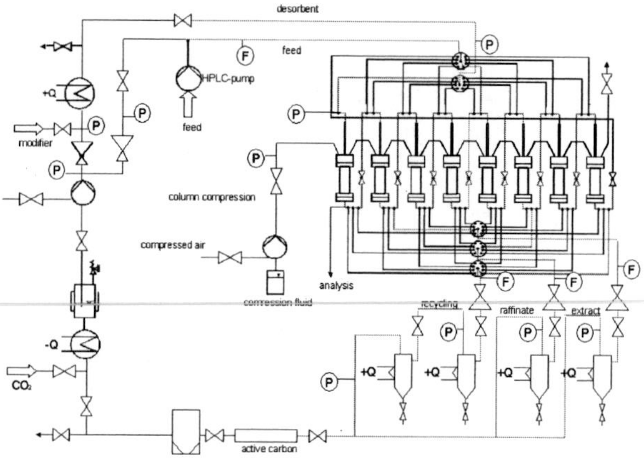

Figure 2.24: Plant scheme

2.7 Example processes

Within this thesis, two separations are considered as application examples: the separation of R–/S–Ibuprofen (Peper et al., 2002) and the separation of α–/δ–Tocopherol (Peper et al., 2003). Both separations were performed in the SMB plant of the Institute of Thermal Separation Sciences at Technical University Hamburg–Harburg (Figure 2.25, (Depta, 1999; Depta et al., 1999)). The plant realises the Supercritical Fluid Chromatography (SFC) using a mixture of supercritical carbon dioxide, to which a modifier is added, as a solvent. Using supercritical fluids as solvent offers many advantages (Johannsen, 2004): Gas–like viscosities enable high flow rates at low pressure drop. The higher diffusion as compared to liquids results in better efficiencies of the separation columns. Additionally, the physical properties of the supercritical fluids can easily be varied by changing the pressure or the temperature or both. Reducing the pressure reduces the solubility which allows to easily recover the product components. The gaseous solvent can be recycled without additional cleaning steps.

The following paragraphs describe the details of the plant and the example separations to which the herein considered application processes are referred.

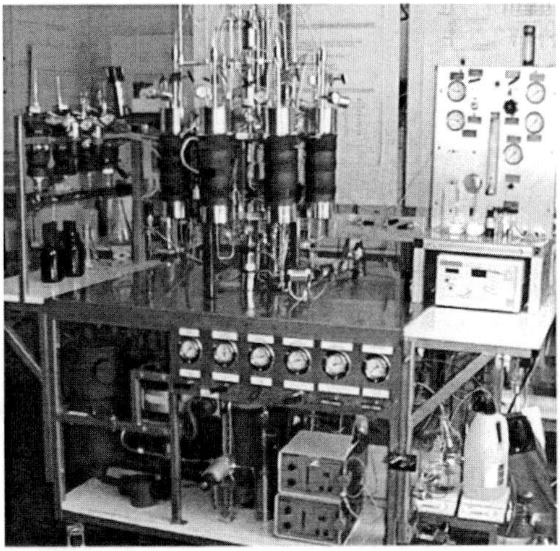

Figure 2.25: SFC–SMB plant of the Institute of Thermal Separation Sciences, Technical University Hamburg–Harburg

SMB plant. The SFC SMB plant has the following properties:

- ◇ eight separation columns,
- ◇ open or closed recycling loop
- ◇ one sample loop, which has a fixed position with respect to the separation columns, for discrete–time concentration measurements,
- ◇ manual control of all process parameters via an operating system,
- ◇ Rheonic flow meters to record the external flow rates $\dot{m}_A$, $\dot{m}_{A+B}$, $\dot{m}_B$ and $\dot{m}_R$ and
- ◇ a pressure sensor to record the pressure profile in the SMB sections.

Figure 2.24 shows the scheme of the SFC plant. The left part of the figure shows the solvent recycling unit and the solvent (desorbent) and feed injection. The unit allows for the injection of a modifier to adjust the solubility of the solvent. The upper right part of the figure shows the circle of separation columns with the column interconnections and the rotating valves, which realise the port switching. The sample loop (analysis) for the determination of the internal concentration profile is located between the very left column and the next column to the right. The lower right part shows the cyclones in which the expansion of the product outlet streams takes place for product recovery.

Example separation problems. Two different example separations are considered. The first refers to the separation of the $S(+)$- and $R(-)$–enantiomers of Ibuprofen on a chiral stationary phase. Ibuprofen is a pain reductant. The $S(+)$–enantiomer has a significantly higher pharmacological efficiency and less side effects than the $R(-)$–enantiomer.

The $S(+)$- and $R(-)$–enantiomers of Ibuprofen show Langmuir–like adsorption behaviour. With respect to the propagation of the concentration profiles in a single–column or an SMB process, this adsorption behaviour leads to a fast propagating and steep adsorption wave front compared to the desorption wave front. Figure 2.26 shows the numerical simulation of two example chromatograms of the $R(-)$–enantiomer. Both chromatograms are simulated with the same injection duration. One chromatogram is simulated with a high injection concentration to show the Langmuir–like behaviour (higher peak). The other one is simulated with a low injection concentration (lower peak).

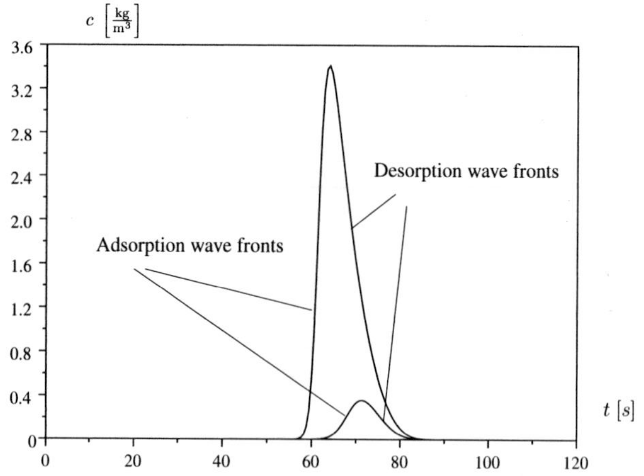

Figure 2.26: Chromatogram of $R(+)$-Ibuprofen with high and low injection concentrations

The second example refers to the separation of α- and δ-Tocopherol on a modified silicagel (Kromasil Si 60–10). In the group of vitamin E, α-Tocopherol has the highest vitamin activity. Its antioxidative properties lead to a stabilisation against other hormones and enzymes. In the technical production, all Tocopherol isomers are formed. The separation allows to obtain the most valuable component, which is α-Tocopherol.

α– and δ–Tocopherol show a contrary adsorption behaviour as compared to Ibuprofen, i.e. the Anti–Langmuir–like adsorption. This means that concentration profiles with high peak concentrations show a steeper and slower propagating desorption wave front compared to the adsorption wave front. Figure 2.27 shows two chromatograms of α–Tocopherol, one with a high and one with a low injection concentration.

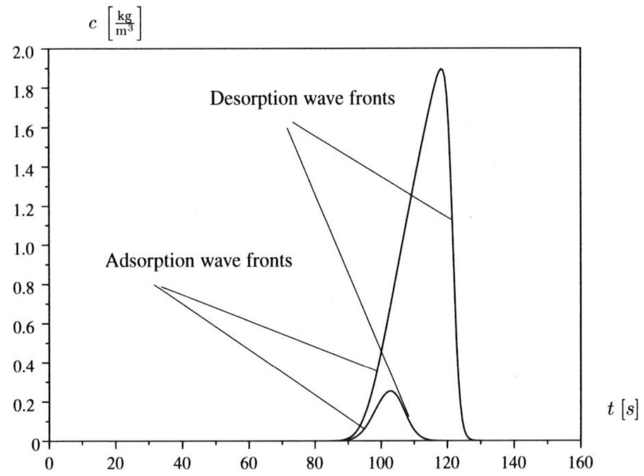

Figure 2.27: Chromatogram of α-Tocopherol with high and low injection concentrations

The process parameters as well as the parameters of the isotherms and the nominal operation point for the two example separations are given in the Appendix B.

Reference processes. Throughout this thesis, the two separation examples are used as a basis for the investigation of the dynamical and the stationary process behaviour and the identification of the process dynamics and serve as an application example for the developed observation and control principles. For the application of the methods it is assumed that the considered SMB plant can be configured as a six column plant and as a four column VARICOL plant.

Within this thesis the following assumptions are made with respect to the behaviour of the example processes:

- spatially and temporally constant isotherm parameters,
- temporally constant pressure profile in the SMB or VARICOL sections,
- nonlinear adsorption which is described by the cubic Hill isotherm,
- negligible extra–column volume differences of the column interconnections,
- negligible differences of the package porosities in the columns,
- disturbance of the processes by a temporal drift in the package porosity and a step of the feed inlet concentration and
- description of the mass exchange and mass transport by the Equilibrium–Dispersive–Model.

Chapter 3

Simulated counterflow chromatography modelling

The derivation of a physical SCC process model is presented in this chapter. The hybrid nature of the processes is explicitly considered. The models of the continuous parts are derived based on a fluid dynamical single–column model. For the TMB, a fluid dynamical model is derived and the stationary solution is given explicitly. The discrete parts of the SMB and the VARICOL are modelled based on a modelling approach for discrete–event systems. The discrete dynamics is represented using state transition and output relations as well as automaton graphs. The separate modelling allows for an isolated analysis of the continuous and the discrete dynamics with respect to the model simplification for the observation and control of SCC processes. The interaction of the continuous and the discrete parts is modelled and studied using the concept of hybrid automata.

3.1 Modelling concept

3.1.1 Hybrid dynamical systems

SMB and VARICOL processes are dynamical systems which encounter both continuous–variable and discrete–event dynamics. These types of systems are referred to as hybrid dynamical systems or briefly hybrid systems. The continuous and the discrete dynamics of SCC processes are each described by a continuous–variable and a discrete–event subsystem, which are also referred to as the continuous and the discrete SCC subsystem in the following. The characteristic issue of the resulting hybrid system is the interaction between the two subsystems (Figure 3.1).

The *discrete SCC subsystem* describes the *port switching*. The discrete input v is the vector of the port switching signals. The discrete output w is the vector of the port positions with respect to the separation columns. The *injector*, which in a physical sense describes the *action of the port switching valves* within the SCC plant setup, assigns to each separation column an internal fluid flow rate according to the actual port positions. It converts the discrete output w into the

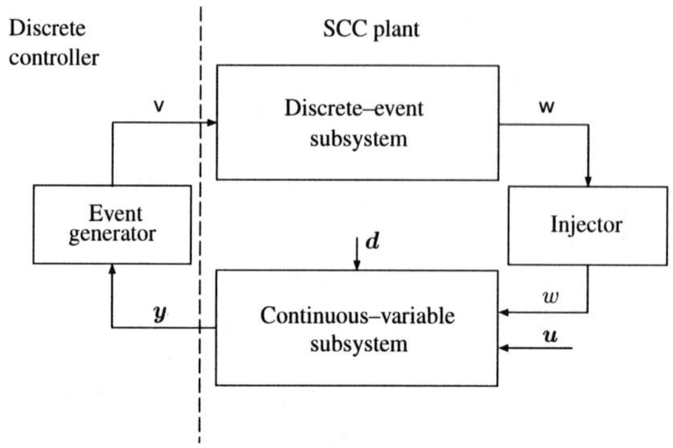

Figure 3.1: Hybrid system of SCC processes

continuous–variable input w which is the vector of the fluid flows in the n_c separation columns. The *continuous SCC subsystem* describes the *fluid dynamical process in the circle of separation columns*. It has, in addition to w, two further input variables: the continuous vector $\boldsymbol{u}$ of the internal fluid flow rates and the continuous vector $\boldsymbol{d}$ of the feed inlet concentrations and the adsorbent package porosity. The output variable $\boldsymbol{y}$ can either be a vector of the continuously measured wave front concentrations or the continuous local time counter τ or τ_S.

An SCC plant consists of the discrete and the continuous SCC subsystems and of the injector (see Figure 3.1). The process performs well for a suitable choice of constant continuous input signals $\boldsymbol{u}$ and $\boldsymbol{d}$ only if the discrete input v is not constant in time. A temporally constant v means that no port switching takes place and, hence, no simulation of the counterflow between the adsorbent and the fluid is performed. A repeated port switching triggered by v is necessary to obtain a well operating SCC plant. The discrete input v is generated by the *event generator* from the continuous output signal $\boldsymbol{y}$. Because the event generator is not part of the SCC plant it has to be designed as a discrete controller such that it applies either a time–invariant or a time–varying switching pattern to the discrete subsystem via v.

The following three paragraphs introduce the model notation of the continuous and discrete SCC subsystems and the hybrid system.

Continuous–variable subsystem. The dynamical behaviour of the continuous–variable subsystem of SCC processes is described by a set of differential equations

$$\dot{\boldsymbol{x}} = \boldsymbol{f}(\boldsymbol{x}(t,z), \boldsymbol{u}(t), \boldsymbol{d}(t), w(t)) \,, \ \boldsymbol{x}(0,z) = \boldsymbol{x}_0(z) \,, \tag{3.1}$$

where $\boldsymbol{f}$ is a given vector operator $\boldsymbol{f}\ :\ \mathbb{R}^{n_x} \times \mathbb{R}^{n_u} \times \mathbb{R}^{n_d} \times \mathbb{R}^{n_w} \mapsto \mathbb{R}^{n_x}$. $\boldsymbol{x}(t,z) \in \mathbb{R}^{n_x}$ is the distributed state variable and $\boldsymbol{x}_0(z)$ is the initial state, $\boldsymbol{u}(t) \in \mathbb{R}^{n_u}$ is the continuous control input and $\boldsymbol{d}(t) \in \mathbb{R}^{n_d}$ is the disturbance input. $w(t) \in \mathbb{R}^{n_w}$ is a continuous signal vector which is generated from the output of the discrete–event subsystem. The dimensions of the respective variables are n_x, n_u, n_d and n_w. The independent variables are the global time $t \in \mathbb{R}_0^+$ and the spatial coordinate $z \in [z_0, z_1]$.

The measured output variable $\boldsymbol{y}(t) \in \mathbb{R}^{n_y}$ is determined by the output equation

$$\boldsymbol{y}(t) = \boldsymbol{g}(\boldsymbol{x}(t,z), \boldsymbol{u}(t), \boldsymbol{d}(t), w(t))\,, \tag{3.2}$$

where $\boldsymbol{g}$ is a given vector operator $\boldsymbol{g}\ :\ \mathbb{R}^{n_x} \times \mathbb{R}^{n_u} \times \mathbb{R}^{n_d} \times \mathbb{R}^{n_w} \mapsto \mathbb{R}^{n_y}$. n_y is the dimension of the continuous output $\boldsymbol{y}$. Figure 3.2 shows the block diagram of the continuous SCC subsystem.

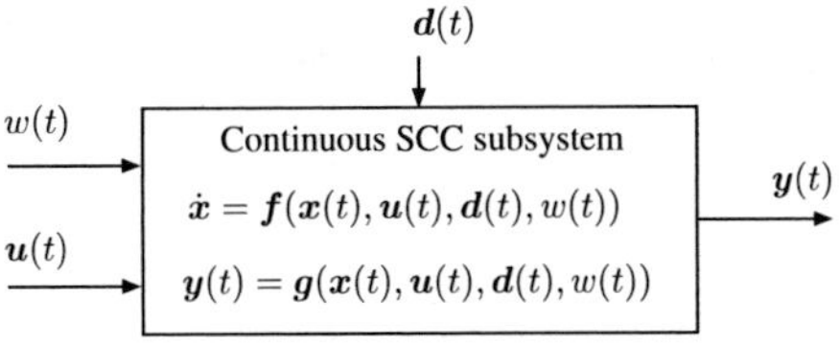

Figure 3.2: Continuous–variable subsystem of SCC processes

Discrete–event subsystem. The dynamical behaviour of the discrete–event subsystem of SCC processes is described by the deterministic state transition relation $G\ :\ \mathcal{N}_{\mathsf{z}}^{n_{\mathsf{z}}} \times \mathcal{N}_{\mathsf{v}}^{n_{\mathsf{v}}} \mapsto \mathcal{N}_{\mathsf{z}}^{n_{\mathsf{z}}}$ and the deterministic output relation $H\ :\ \mathcal{N}_{\mathsf{z}}^{n_{\mathsf{z}}} \times \mathcal{N}_{\mathsf{v}}^{n_{\mathsf{v}}} \mapsto \mathcal{N}_{\mathsf{w}}^{n_{\mathsf{w}}}$ with

$$\begin{aligned} \mathsf{z}(k+1) &= G(\mathsf{z}(k), \mathsf{v}(k)) \\ \mathsf{w}(k) &= H(\mathsf{z}(k), \mathsf{v}(k))\,, \end{aligned} \tag{3.3}$$

where $\mathsf{v}(k) \in \mathcal{N}_{\mathsf{v}}^{n_{\mathsf{v}}}$ is the discrete input variable, $\mathsf{z}(k) \in \mathcal{N}_{\mathsf{z}}^{n_{\mathsf{z}}}$ is the discrete state variable, with the initial state $\mathsf{z}(0) = \mathsf{z}_0$, and $\mathsf{w}(k) \in \mathcal{N}_{\mathsf{w}}^{n_{\mathsf{w}}}$ is the discrete output variable. In the following, the successor state $\mathsf{z}(k+1) \in \mathcal{N}_{\mathsf{z}}^{n_{\mathsf{z}}}$ is also denoted by $\mathsf{z}'(k)$ or simply z'. The input, state and output variables are vectors of the dimensions n_{v}, n_{z} and n_{w} and the vector elements are symbols of the sets $\mathcal{N}_{\mathsf{v}}$, $\mathcal{N}_{\mathsf{z}}$ and $\mathcal{N}_{\mathsf{w}}$. The independent variable is the counter $k \in \mathbb{N}_0$ of the state transition. Figure 3.3 shows the block diagram of the discrete subsystem.

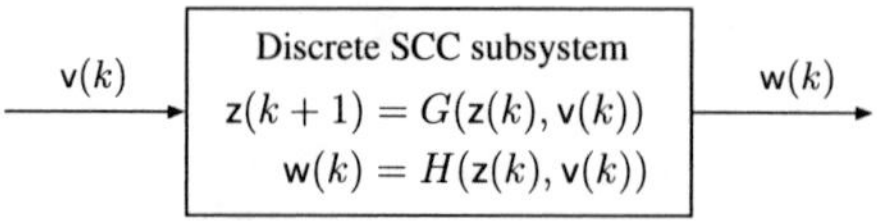

Figure 3.3: Discrete–event subsystem of SCC processes

Hybrid system. The hybrid nature of SCC processes arises from the coupling of the two subsystems. The discrete output $\mathsf{w}(k)$ is injected to the continuous subsystem via the injector $I : \mathcal{N}_{\mathsf{w}}^{n_{\mathsf{w}}} \mapsto \mathbb{R}^{n_{\mathsf{w}}}$ with

$$w(t) = I(\mathsf{w}(k)) \,. \tag{3.4}$$

The continuous input variable w is also denoted as a function $w = w(\mathsf{w}(k))$. The injector of SCC processes is given by the physical setup of the modelled plant. The event generator with $E : \mathbb{R}^{n_y} \mapsto \mathcal{N}_{\mathsf{v}}^{n_{\mathsf{v}}}$ generates the discrete input $\mathsf{v}(k)$ from the continuous output $\boldsymbol{y}(t)$

$$\mathsf{v}(k) = E(\boldsymbol{y}(t)) \tag{3.5}$$

and thereby connects the output of the continuous subsystem to the input of the discrete subsystem. The event generator of SCC processes is not prescribed by the physical plant setup and has to be designed as a discrete controller.

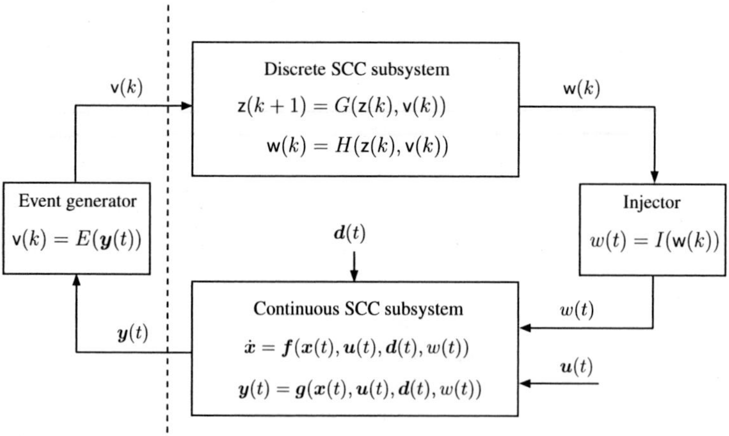

Figure 3.4: Hybrid dynamics of SCC processes

Figure 3.4 shows a block diagram representation of the hybrid dynamical system of SCC processes. The system has the following properties:

- ◇ The discrete–event subsystem is deterministic.
- ◇ The continuous–variable subsystem is a distributed–parameter system.
- ◇ The subsystems are interconnected by the injector and the event generator.
- ◇ The event generator is the discrete controller.
- ◇ The continuous–variable subsystem has a free continuous control input $\boldsymbol{u}$ and a continuous disturbance input $\boldsymbol{d}$.

3.1.2 Modelling procedure

The aim of the modelling is to provide a model of SCC processes which allows for the separate and detailed analysis of

- ◇ the continuous,
- ◇ the discrete and
- ◇ the combined continuous and discrete

dynamics as a basis for the model reduction and simplification. Therefore, the continuous and the discrete subsystems of SCC processes are modelled separately. The overall system dynamics is modelled by connecting the two subsystems. To be able to perform this modelling step it has to be regarded that each of the subsystem models encounter the interconnections within the hybrid system model. The first approach to this modelling concept was published in (Kleinert and Lunze, 2002) and shows an application to SMB processes with four separation columns. Within this thesis, the approach is generalised and applied to the modelling of SCC processes. The following paragraphs describe the steps of the modelling procedure in detail.

Continuous–variable subsystem. The continuous dynamics of SCC processes are governed by the fluid dynamics which encounter mass transport and mass exchange in the separation columns of the process. Because SCC processes consist of several single separation columns, the single–column chromatographic separation is modelled first in Section 3.2. First principles modelling of the fluid dynamics yields the Convection–Diffusion–Equation, which describes the dynamics of the distributed concentration profiles in a separation column. The presented modelling approach is well known and widely applied for the description of chromatographic separations in the context of SCC simulation and control (Seidel-Morgenstern, 1995; Dünnebier et al., 1998; Ludemann-Homburger et al., 2000; Giese, 2002; Hanisch, 2002; Erdem et al., 2004; Toumi, 2005). The derivation is presented here to show on which kind of models the subsequent system modelling and analysis approaches are based on.

The behaviour of the TMB process shows many similarities to that of SMB and VARICOL processes, because it is also based on the counterflow principle. Especially the shape of the concentration profiles is similar to that of SCC processes. This similarity is used for a model reduction to solve the SMB wave front observation task. To obtain a description of the shape of the TMB concentration profiles, a physical model is derived based on first principles modelling of the mass transport and exchange of the counterflow. The analytic solution of the dynamical transition and the explicit analytic expression for the stationary concentration profiles are derived assuming a closed–loop separation column of the TMB. The modelling approach is presented in Section 3.3. The concept was published in (Kleinert and Lunze, 2003, 2005) and was applied to the derivation of a wave front model.

SMB and VARICOL processes have principally the same type of continuous subsystems. The model of this subsystem is obtained from single–column models regarding the node balances of the column interconnections. This approach is presented in Section 3.4. As the presented model for the single–column separation, the modelling of the continuous subsystems of SCC processes is well known in the chromatography community. A new approach of visualising the continuous SCC subsystems by means of a block diagram is introduced considering varying column characteristics and extra–column volumes of the column interconnections. This representation shows how the model parameters of the column models change in case of a port switching.

Discrete–event subsystem. The discrete SCC subsystem describes the change of the port positions if a switching signal for the respective port is applied. Because of the setup of SCC plants this subsystem is deterministic. The discrete subsystem of SMB processes describes the synchronous switching of the ports, which leads to a subsystem representation of low complexity. The discrete subsystem of VARICOL processes is more complex due to the multiple possibilities of port position combinations on the circle of separation columns. The modelling of the discrete–event dynamics is presented in Section 3.5.

Hybrid dynamics of SCC processes. Based on the modelling of the continuous–variable and the discrete–event subsystems, the overall system behaviour is represented using the concept of the hybrid automaton in Section 3.6. The properties of the overall system model of SMB and VARICOL processes are investigated. The analysis shows how the subsystems of SCC processes interact.

3.2 Single–column separations

3.2.1 Fluid dynamical modelling

Mass transport and mass exchange. The central components of TMB, SMB and VARICOL processes are tubes, the so–called separation columns, which contain the adsorbent in a fixed porous bed or, in case of the TMB process, in a moving porous bed. The bed of adsorbent consists of globular, porous particles. Inside of the column, the fluid stream flows through this bed and thereby gets in contact with the adsorbent surface (Figure 3.5).

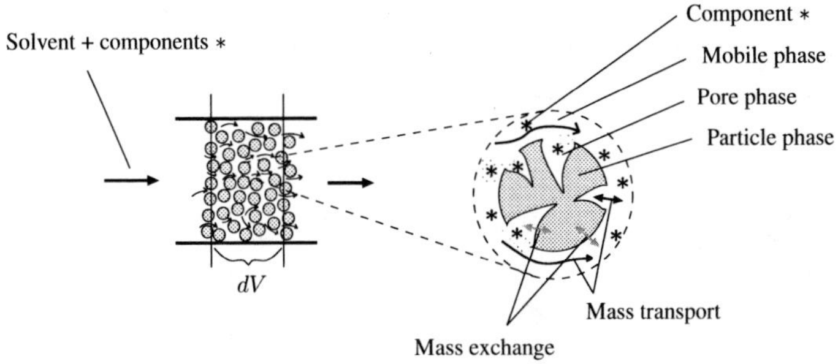

Figure 3.5: Mass exchange between the fluid and the adsorbent

The fluid stream, which flows through the porous bed, is called mobile phase and carries the diluted components of the mixture indicated by "$*$". In case of the SMB or the VARICOL, the porous bed is fixed whereas in case of the TMB, the porous bed propagates in the opposite direction of the fluid flow. The fluid inside of the pores is referred to as the pore phase, and the particle surface is referred to as the particle phase.

The following types of mass transport and mass exchange take place in the mobile and particle phase:

1. The diluted mixture components are carried by the moving mobile phase. This mass transport is called the *convective mass transport*.

2. Due to a concentration gradient a *diffusive mass transport* occurs in the mobile phase.

3. The contact of the fluid to the particle surface yields a *mass exchange* between the fluid (mobile phase *and* pore phase) and the particle phase. The driving force for the mass exchange is the adsorption and desorption tendency of the considered components $*$.

4. A further mass transport takes place between the mobile phase and the pore phase. It is a diffusive transport due to a local concentration gradient.

5. In case of large particle diameters a considerable concentration gradient occurs inside of the pore phase and, hence, also on the inner particle surface. This gradient leads to a diffusive mass transport in the pore phase and the particle phase.

Dynamical models of this mass transport and mass exchange system are derived by fluid dynamical modelling approaches using mass balances of infinitesimally small volume elements dV of the column interior (see (Spieker, 2000; Kleinert, 2002)). Several models are published in literature. Depending on the mass transport resistance in the solvent and on the adsorbent, different combinations of the previously described mass transport and mass exchange mechanisms are considered (Dünnebier and Klatt, 2000). For the modelling approach which is presented in this thesis, the following mass transport and mass exchange is considered:

- ◇ convective and diffusive mass transport in the mobile phase,
- ◇ local adsorption equilibrium between the particle phase and the fluid in the balance volume dV,
- ◇ negligible mass transport resistance between the mobile phase and the pore phase and, hence, no concentration gradient between the mobile and the pore phase,
- ◇ mass transport by the propagation of the adsorbent in case of the TMB process,
- ◇ negligible radial concentration gradients inside of the separation column.

Model derivation. Based on these assumptions, the fluid dynamical modelling of the mass transport and mass exchange in a single–column can be performed using the partial mass balance of an infinitesimally small volume element dV of the column interior. According to the homogenisation principle, it is assumed that the overall geometrical relation between the adsorbent and the mobile phase is homogeneous in the whole volume of the bed of adsorbent and also applies to an infinitesimally small balance volume dV. Then, the balance volume can be reduced as shown in Figure 3.6: To derive the mass balance, the local balance volume dV is divided into one part dV_m which refers to the mobile phase including the pore phase, and one part dV_p which refers to the particle phase such that the balance volume shown in Figure 3.6 (b) is obtained from the physical geometry in Figure 3.6 (a). A local adsorption equilibrium is assumed between the fluid volume dV_m and the particle phase dV_p. For the derivation of the mass balance, the partial mass flow rate of the mixture component $*$ is considered. $\dot{m}_{in,*,z}$ describes the partial mass flow which is entering the balance volume at the position z. $\dot{m}_{out,*,z}$ represents the partial mass flow which leaves the volume element at the position $z+dz$. In a single–column separation no moving adsorbent has to be considered. Hence, $\dot{m}_{in,*,z}$ and $\dot{m}_{out,*,z}$ are composed of the convective and diffusive partial mass flow rates $\dot{m}_{C,*,z}$ and $\dot{m}_{D,*,z}$

$$\begin{aligned} \dot{m}_{in,*,z} &= \dot{m}_{C,*,z} + \dot{m}_{D,*,z} \\ \dot{m}_{out,*,z} &= \dot{m}_{C,*,z+dz} + \dot{m}_{D,*,z+dz} \, , \end{aligned}$$

and the balance volume as shown in Figure 3.6 (c) is obtained.

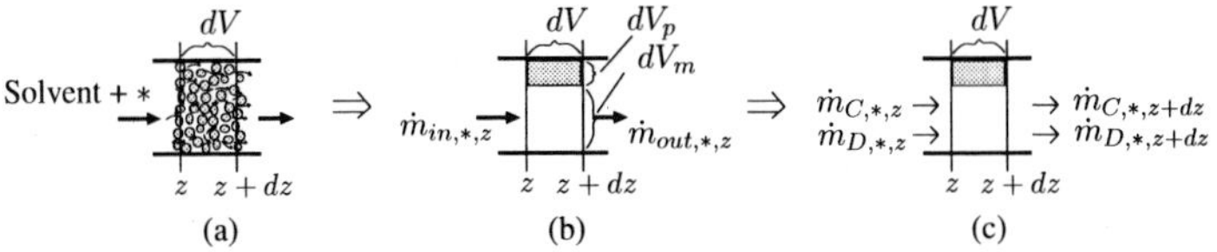

Figure 3.6: Balance volume dV for the fluid dynamical modelling of the single–column chromatography

In the following, the physical model for the single–column chromatographic separation is derived in five steps:

Step 1. The basic step of deriving the partial mass balance of the component $*$ for dV is based on the law of mass conservation. The temporal variation of the partial mass content m_* of the balance volume is determined by the sum of the partial mass flow rates entering and leaving the volume element:

$$\begin{aligned} \left.\frac{\partial m_*}{\partial t}\right|_z &= \dot{m}_{in,*,z} - \dot{m}_{out,*,z} \\ &= \dot{m}_{C,*,z} + \dot{m}_{D,*,z} - \dot{m}_{C,*,z+dz} - \dot{m}_{D,*,z+dz} \, . \end{aligned} \tag{3.6}$$

Step 2. The mass flow rates are a function of z. The mass flow rates at the position $z + dz$ are determined by applying the Taylor expansion to the mass flow rates at the position z:

$$\begin{aligned} \dot{m}_{C,*,z+dz} &= \dot{m}_{C,*,z} + \frac{\partial \dot{m}_{C,*,z}}{\partial z} dz + \frac{\partial^2 \dot{m}_{C,*,z}}{\partial z^2} \frac{dz^2}{2!} + \ldots \\ \dot{m}_{D,*,z+dz} &= \dot{m}_{D,*,z} + \frac{\partial \dot{m}_{D,*,z}}{\partial z} dz + \frac{\partial^2 \dot{m}_{D,*,z}}{\partial z^2} \frac{dz^2}{2!} + \ldots \, . \end{aligned} \tag{3.7}$$

Applying Equation (3.7) to Equation (3.6) yields the expression

$$\left.\frac{\partial m_*}{\partial t}\right|_z = -\frac{\partial \dot{m}_{C,*,z}}{\partial z}\,dz - \left(\frac{\partial^2 \dot{m}_{C,*,z}}{\partial z^2}\frac{dz^2}{2!} + \ldots\right) - \frac{\partial \dot{m}_{D,*,z}}{\partial z}\,dz - \left(\frac{\partial \dot{m}_{D,*,z}}{\partial z}\frac{dz^2}{2!} + \ldots\right) \quad (3.8)$$

for the temporal change of the partial mass content m_* in dV.

Step 3. The partial mass content m_* in the volume $dV = A_c\,dz$, where A_c is the column cross section area, depends upon the local component concentration $c_*(t,z)$ in the solvent and $q_*(t,z)$ on the adsorbent surface:

$$\begin{aligned} m_* &= \varepsilon\,dV\,c_*(t,z) + (1-\varepsilon)\,dV\,q_*(t,z) \\ &= (\varepsilon\,A_c\,c_*(t,z) + (1-\varepsilon)\,A_c\,q_*(t,z))\,dz\,. \end{aligned} \quad (3.9)$$

ε is the package porosity of the adsorbent, which is defined as the fraction of the interparticle volume V_{int} (i.e. that of the mobile phase in the package) and the volume V_c of the adsorbent package plus mobile phase (which corresponds to the inner column volume, if a time–invariant porosity is considered):

$$\varepsilon = \frac{V_{int}}{V_c}\,.$$

$c_*(t,z)$ is the component concentration in the solvent and $q_*(t,z)$ the concentration on the adsorbent surface. It is assumed that the geometric parameters A_c and ε are spatially and temporally independent and are homogeneous with respect to the size of the balance volume dV, i.e. the values are independent of the incremental length dz. The concentrations $c_*(t,z)$ and $q_*(t,z)$ are considered as the only spatially and temporally dependent variables in Equation (3.9).

The partial mass flow rates $\dot{m}_{C,*,z}$ and $\dot{m}_{D,*,z}$ are expressed using the component concentrations by

$$\begin{aligned} \dot{m}_{C,*,z} &= u\,\varepsilon\,A_c\,c_*(t,z) \\ \dot{m}_{D,*,z} &= -D\,\varepsilon\,A_c\,\tfrac{\partial c_*(t,z)}{\partial z}\,, \end{aligned} \quad (3.10)$$

where u is the flow velocity of the internal fluid and D is the diffusion coefficient. For the modelling, both parameters are considered as time–invariant and spatially independent.

Step 4. It is assumed that for all z a local adsorption equilibrium between the mobile phase and the adsorbent surface occurs. The equilibrium relation between $c_*(t,z)$ and $q_*(t,z)$ is described by the adsorption isotherm f_{Eq} according to Equation (2.1):

$$\boxed{q_*(t,z) = f_{Eq}\left(c_*(t,z)\right) .} \tag{3.11}$$

Step 5. Applying Equations (3.9), (3.10) and (3.11) to Equation (3.8) and considering the limit of $dz \rightarrow 0$ yields the partial differential equation

$$\frac{\partial c_*(t,z)}{\partial t} + F\,\frac{\partial q_*(t,z)}{\partial t} = -u\,\frac{\partial c_*(t,z)}{\partial z} + D\,\frac{\partial^2 , c_*(t,z)}{\partial z^2} , \tag{3.12}$$

where $F = \frac{1-\varepsilon}{\varepsilon}$. Applying the adsorption isotherm (3.11) to Equation (3.12) yields

$$\boxed{\frac{\partial c_*(t,z)}{\partial t} = F'_* \left(\underbrace{-u\,\frac{\partial c_*(t,z)}{\partial z}}_{\text{Convective term}} + \underbrace{D\,\frac{\partial^2 c_*(t,z)}{\partial z^2}}_{\text{Diffusive term}} \right) , \quad z \in [z_0, z_1]} \tag{3.13}$$

for $t \geq 0$. Equation (3.13) is the Convection–Diffusion–Equation. In literature, this model is also called the Equilibrium–Dispersive–Model of chromatographic separations (Spieker, 2000; Dünnebier and Klatt, 2000). The right–hand side of the equation describes the temporal derivative of the local concentration $c_*(t,z)$ of $*$ in the mobile phase. It includes two terms which describe the convective and the diffusive mass transport in the mobile phase. The diffusive term is also said to describe the dispersion in the mobile phase, which encounters the diffusion and remixing effects caused by wall friction and turbulent flow in the mobile phase. Both the convective and the diffusive term are weighted by a factor

$$\boxed{F'_* = \frac{1}{1+F\,\frac{\partial q_*(t,z)}{\partial c_*(t,z)}} ,} \tag{3.14}$$

which is constant for a linear adsorption isotherm (3.11). The spatial coordinate z in Equation (3.13) is defined over the range $z \in [z_0, z_1]$.

Because Equation (3.13) encounters the first temporal and the first and the second spatial derivative of c_*, one initial condition and two boundary conditions are necessary to provide a solution of the model equation.

Initial condition. The initial condition for the Equilibrium–Dispersive–Model (3.13) is

$$\boxed{c_*(0,z) = c_{0,*}(z)\,.} \tag{3.15}$$

Boundary conditions. For the derivation of the boundary conditions, different kinds of approaches can be found in literature depending on the physical setup of the considered separation column. A common approach is to apply the Danckwerts boundary conditions (Danckwerts, 1953), which are derived assuming a negligible diffusion at the inlet and the outlet of the separation column. Considering these boundary conditions, an explicit solution of the Equilibrium–Dispersive–Model can be derived assuming linear adsorption (Lapidus and Amundson, 1952). The boundary conditions imply a horizontal slope of the concentration profile at the column inlet and outlet. This assumption holds for an infinitesimal long column with $z \in [-\infty, \infty]$, which, however, is not reasonable for SCC processes, where commonly short columns are used. Hence, the following mixed boundary conditions are used here, which are derived from local mass balances at the column inlet $z = 0$ and the $z = L$:

$$\begin{aligned} \left(u\, c_{*,in}(t) - D\,\tfrac{\partial c_*}{\partial z}\right)\Big|_{z=0^-} &= \left(u\, c_*(t,z) - D\,\tfrac{\partial c_*}{\partial z}\right)\Big|_{z=0^+} \\ \left(u\, c_{*,out}(t) - D\,\tfrac{\partial c_*}{\partial z}\right)\Big|_{z=L^+} &= \left(u\, c_*(t,z) - D\,\tfrac{\partial c_*}{\partial z}\right)\Big|_{z=L^-}\,. \end{aligned} \tag{3.16}$$

The concentration of the component $*$ in the fluid stream which is entering or leaving the column, respectively, is indicated by $c_{*,in}(t)$ and $c_{*,out}(t)$. These concentrations may be prescribed by the connections attached to the considered separation column as it is the case in the circle of separation columns in SCC processes, or in the direct connections of the sections of the TMB process (see Section 3.3.3). The diffusion which occurs in the tubes connected to the inlet and outlet of the column is small compared to the dispersion inside of the separation column and can be neglected. Hence, $D|_{z=0^-} = D|_{z=L^+} = 0$ and the following mixed boundary conditions are obtained from Equation (3.16):

$$c_{*,in}(t) = c_*(t,z) - \frac{D}{u}\frac{\partial c_*}{\partial z}\,, \qquad z = 0 \tag{3.17a}$$

$$c_{*,out}(t) = c_*(t,z) - \frac{D}{u}\frac{\partial c_*}{\partial z}\,, \qquad z = L\,. \tag{3.17b}$$

The boundary conditions (3.17a) and (3.17b) were used for the implementation of the numerical solver of the Equilibrium–Dispersive–Model in the CSep toolbox and for the derivation of the analytic solution of the physical TMB model.

3.2.2 Solution of the fluid dynamical model

If a linear adsorption is assumed, the factor F_*' is a constant. Then, the analytic solution of the fluid dynamical model can be derived, which is presented in this section.

In the following it is assumed that z is normalised such that $z_0 = 0$ and $z_1 = 1$. The general solution $c_*(t,z)$ for Equation (3.13) is obtained by the superposition of the dynamical solution and the stationary solution:

$$c_*(t,z) = c_{dyn,*}(t,z) + c_{stat,*}(z)\,.$$

The dynamical solution $c_{dyn,*}(t,z)$ is determined by the initial condition $c_*(0,t)$ (denoted by "ic") and the boundary conditions $c_{*,in}(t)$ and $c_{*,out}(t)$ (denoted by "bc") :

$$c_{dyn,*}(t,z) = c_{dyn,*}(t,z)\big|_{\text{ic}} + c_{dyn,*}(t,z)\big|_{\text{bc}}\,.$$

The stationary solution $c_{stat,*}(z)$ is different from zero if $c_{*,in}(t)$, or $c_{*,out}(t)$, or both, have constant terms $\bar{c}_{*,in} \neq 0$ or $\bar{c}_{*,out} \neq 0$, respectively.

The dynamical solution is obtained from the application of the separation principle:

$$c_{dyn,*}(t,z) = \sum_{i=1}^{\infty} g_{*,i}(t) \cdot h_{*,i}(z)\,. \tag{3.18}$$

By applying Equation (3.18) to (3.13) the functions $g_{*,i}(t)$ and $h_{*,i}(z)$ are obtained

$$
\begin{aligned}
g_{*,i}(t) &= \mathrm{e}^{\lambda_{*,i}\,t} \\
h_{*,i}(z) &= p_{1,*,i} \cdot \mathrm{e}^{q_{1,*,i}\,z} + p_{2,*,i} \cdot \mathrm{e}^{q_{2,*,i}\,z} ,
\end{aligned}
$$

with

$$
\begin{aligned}
q_{1,*,i} &= \tfrac{u}{2\,D} + \sqrt{\tfrac{u^2}{4\,D^2} + \tfrac{\lambda_{*,i}}{F'_*\,D}} \\
q_{2,*,i} &= \tfrac{u}{2\,D} - \sqrt{\tfrac{u^2}{4\,D^2} + \tfrac{\lambda_{*,i}}{F'_*\,D}} .
\end{aligned}
$$

This result shows that $q_{1,*,i}$ and $q_{2,*,i}$ depend upon $\lambda_{*,i}$. Therefore, the parameters of Equation (3.18) are $\lambda_{*,i}$, $p_{1,*,i}$ and $p_{2,*,i}$. The complete dynamical solution is given by

$$
\begin{aligned}
c_{dyn,*}(t,z) &= \sum_{i=1}^{\infty} \left(\mathrm{e}^{\lambda_{*,i}\,t} \left(p_{1,*,i} \cdot \mathrm{e}^{q_{1,*,i}\,z} + p_{2,*,i} \cdot \mathrm{e}^{q_{2,*,i}\,z}\right)\right)\Bigg|_{\mathrm{ic}} \\
&+ \sum_{i=1}^{\infty} \left(\mathrm{e}^{\lambda_{*,i}\,t} \left(p_{1,*,i} \cdot \mathrm{e}^{q_{1,*,i}\,z} + p_{2,*,i} \cdot \mathrm{e}^{q_{2,*,i}\,z}\right)\right)\Bigg|_{\mathrm{bc}} .
\end{aligned} \tag{3.19}
$$

Because Equation (3.19) has to fulfil the initial condition (3.15) for $t = 0$ and the boundary conditions (3.17a) and (3.17b) for $z = 0$ and $z = 1$, the parameters $\lambda_{*,i}$, $p_{1,*,i}$ and $p_{2,*,i}$ are determined by the initial condition and the boundary conditions.

The stationary solution is obtained by the solution of Equation (3.13) for $\frac{\partial c_*}{\partial t} = 0$. Then, an ordinary homogeneous second–order differential equation with respect to z is obtained:

$$
\frac{d^2\, c_{stat,*}}{dz^2} - \frac{u}{D}\,\frac{d\,c_{stat,*}}{dz} = 0 .
$$

Because the differential equation contains only the first and the second spatial derivative, the solution yields

$$
c_{stat,*}(z) = r_* \cdot \mathrm{e}^{\frac{u}{D}\,z} + s_* . \tag{3.20}
$$

The constants r_* and s_* are determined by the application the boundary conditions (3.17a) and (3.17b). For the general solution of Equation (3.13)

$$
\begin{aligned}
c_{dyn,*}(t,z) &= \sum_{i=1}^{\infty} \left(\mathrm{e}^{\lambda_{*,i}\,t} \left(p_{1,*,i} \cdot \mathrm{e}^{q_{1,*,i}\,z} + p_{2,*,i} \cdot \mathrm{e}^{q_{2,*,i}\,z}\right)\right)\Bigg|_{\mathrm{ic}} \\
&+ \sum_{i=1}^{\infty} \left(\mathrm{e}^{\lambda_{*,i}\,t} \left(p_{1,*,i} \cdot \mathrm{e}^{q_{1,*,i}\,z} + p_{2,*,i} \cdot \mathrm{e}^{q_{2,*,i}\,z}\right)\right)\Bigg|_{\mathrm{bc}} \\
&+ r_* \cdot \mathrm{e}^{\frac{u}{D}\,z} + s_*
\end{aligned} \tag{3.21}
$$

is obtained.

3.3 True Moving Bed processes

The True Moving Bed process realises the real counterflow between the fluid and the adsorbent in a single separation column. In this section, the derivation of a physical model of the TMB process is presented. The general analytic solution of the model is given and an explicit solution of the stationary TMB concentration profile is derived including an explicit expression for the parameters of the solution.

3.3.1 Fluid dynamical modelling

The physical model of the TMB process is derived according to the derivation steps of the single–column separation model. However, compared to the single–column separation, the partial mass transport by the movement of the adsorbent has to be considered. Figure 3.7 (a), (b) and (c) show how the physical distribution of mobile phase and the particles in dV is reduced to the volumes dV_m and dV_p. Corresponding to the model derivation of the single separation column, a convective and a diffusive partial mass flow $\dot{m}_{C,*}$ and $\dot{m}_{D,*}$ is considered in the fluid. Additionally, the partial mass transport $\dot{m}_{Q,*}$ by the adsorbent is considered, which points in the opposite direction of the fluid flow. Hence, the partial mass flows $\dot{m}_{in,*,z}$ and $\dot{m}_{out,*,z}$, which enter or leave the volume element dV, are determined as follows:

$$\begin{aligned} \dot{m}_{in,*,z} &= \dot{m}_{C,*,z} + \dot{m}_{D,*,z} - \dot{m}_{Q,*,z} \\ \dot{m}_{out,*,z} &= \dot{m}_{C,*,z+dz} + \dot{m}_{D,*,z+dz} - \dot{m}_{Q,*,z+dz} \,. \end{aligned}$$

In the following, the physical TMB model is derived in five steps.

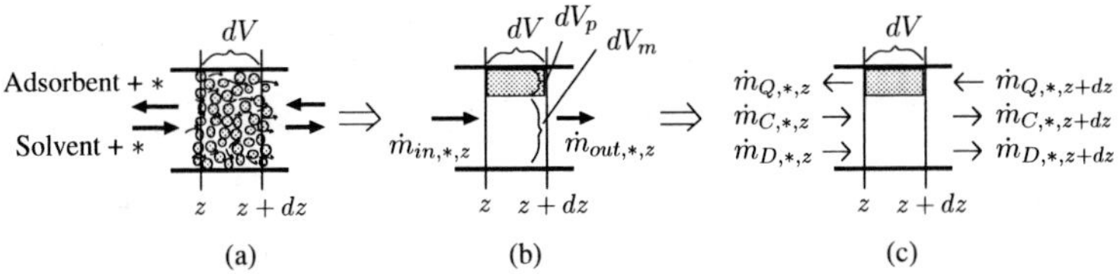

Figure 3.7: Fluid dynamical modelling of the true counterflow chromatographic process

Step 1. The partial mass balance for the component $*$ in dV yields

$$\begin{aligned} \left.\frac{\partial m_*}{\partial t}\right|_z &= \dot{m}_{in,*} - \dot{m}_{out,*} \\ &= \dot{m}_{C,*,z} + \dot{m}_{D,*,z} - \dot{m}_{Q,*,z} \\ & \quad -\dot{m}_{C,*,z+dz} - \dot{m}_{D,*,z+dz} + \dot{m}_{Q,*,z+dz} \;. \end{aligned} \tag{3.22}$$

Step 2. Equation (3.7) is applied to replace $\dot{m}_{C,*,z+dz}$ and $\dot{m}_{D,*,z+dz}$. The flow rate $\dot{m}_{Q,*,z+dz}$ is replaced using the following Taylor expansion:

$$\dot{m}_{Q,*,z+dz} = \dot{m}_{Q,*,z} + \frac{\partial \dot{m}_{Q,*,z}}{\partial z}\, dz + \frac{\partial^2 \dot{m}_{Q,*,z}}{\partial z^2}\frac{dz^2}{2!} + \dots \;. \tag{3.23}$$

Because the higher terms of the Taylor expansion vanish if an infinitesimally small balance volume with $dz \to 0$ is considered, Equation (3.22) is transformed to

$$\left.\frac{\partial m_*}{\partial t}\right|_z = -\frac{\partial \dot{m}_{C,*,z}}{\partial z}\, dz + \frac{\partial \dot{m}_{Q,*,z}}{\partial z}\, dz - \frac{\partial \dot{m}_{D,*,z}}{\partial z}\, dz \tag{3.24}$$

for the further derivation steps.

Step 3. The partial mass content m_* in dV is determined using Equation (3.9):

$$m_* = \left(\varepsilon\, A_c\, c_*(t,z) + (1-\varepsilon)\, A_c\, q_*(t,z)\right) dz \,. \tag{3.25}$$

The flow rates describing the convective and the diffusive mass transport are determined by

$$\begin{aligned} \dot{m}_{C,*,z} &= u_j\, \varepsilon\, A_c\, c_*(t,z) \\ \dot{m}_{Q,*,z} &= u_s\, (1-\varepsilon)\, A_c\, q_*(t,z) \\ \dot{m}_{D,*,z} &= -D\, \varepsilon\, A_c\, \frac{\partial c_*(t,z)}{\partial z}\,, \end{aligned} \tag{3.26}$$

where u_j is the flow velocity of the internal fluid which is piecewise constant in the considered TMB section j. u_s is the adsorbent propagation velocity and D is the diffusion coefficient.

Step 4. The dependence of $c_*(t,z)$ upon $q_*(t,z)$ is determined by the adsorption isotherm

$$q_*(t,z) = f_{Eq,*}\left(c_*(t,z)\right) \,. \tag{3.27}$$

Step 5. From Equation (3.24) the model for the TMB section j can be derived using Equation (3.25) and (3.26):

$$\frac{\partial c_*(t,z)}{\partial t} + F\,\frac{\partial q_*(t,z)}{\partial t} = -v_{*,j}\,\frac{\partial c_*(t,z)}{\partial z} + D\,\frac{\partial^2, c_*(t,z)}{\partial z^2}\,, \tag{3.28}$$

with $F = \frac{1-\varepsilon}{\varepsilon}$. With Equation (3.27), Equation (3.28) is transformed to

$$\boxed{\frac{\partial c_*(t,z)}{\partial t} = F'\left(-v_{*,j}\,\frac{\partial c_*(t,z)}{\partial z} + D\,\frac{\partial^2 c_*(t,z)}{\partial z^2}\right),} \tag{3.29}$$

with

$$\boxed{F' = \frac{1}{1+F\,\frac{\partial f_{Eq,*}(t,z)}{\partial c_*(t,z)}}\,.}$$

Equation (3.29) corresponds to the Equilibrium–Dispersive–Model for the single separation column. The difference to that equation arises from the parameter $v_{*,j}$ which encounters the counterflow between the solvent and the adsorbent. The parameter is called the *relative flow velocity*

$$\boxed{v_{*,j} = u_j - u_s\,F\frac{\partial f_{Eq,*}}{\partial c_*}\,.} \tag{3.30}$$

Complete TMB model. For the derivation of the TMB model, a separation of a mixture with two components A and B and the implementation of the TMB in a separation column which forms a loop is considered (Figure 3.8).

Port	Position	Section	Range
Solvent inlet	$z = 0$	I	$z \in [0,1]$
Outlet of A	$z = 1$	II	$z \in [1,2]$
Feed inlet	$z = 2$	III	$z \in [2,3]$
Outlet of B	$z = 3$	IV	$z \in [3,4]$

Table 3.1: Port positions and section ranges

The TMB process consists of four connected spatial sections $j = I, II, III, IV$ each with different relative flow velocities $v_{A,j}$ and $v_{B,j}$. The positions of the inlet and outlet ports and the spatial

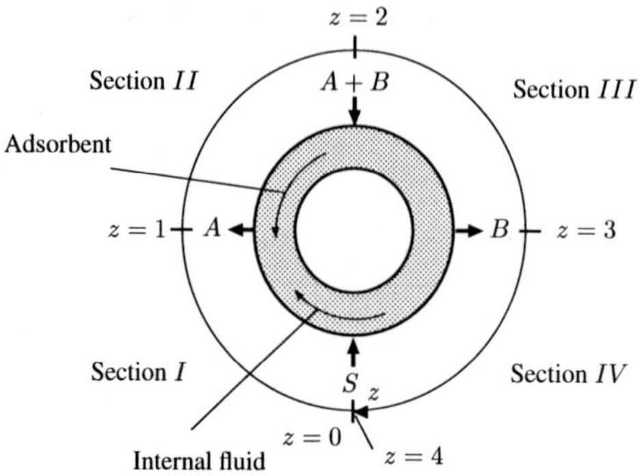

Figure 3.8: TMB process with a closed–loop separation column

ranges of the TMB sections are indicated on the fixed coordinate system z in Figure 3.8 (see also Table 3.1).

To obtain the complete TMB model, Equation (3.29) has to be applied to each TMB section. This yields the following differential equations for the concentration profiles $c_{A,j}(t,z)$ and $c_{B,j}(t,z)$, $j = I, II, III, IV$:

$$
\begin{aligned}
I: \quad z \in [0,1] \quad \frac{\partial c_{A,I}}{\partial t} &= F'_A \left(-v_{A,I} \frac{\partial c_{A,I}}{\partial z} + D \frac{\partial^2 c_{A,I}}{\partial z^2} \right) \\
\frac{\partial c_{B,I}}{\partial t} &= F'_B \left(-v_{B,I} \frac{\partial c_{B,I}}{\partial z} + D \frac{\partial^2 c_{B,I}}{\partial z^2} \right) \\
II: \quad z \in [1,2] \quad \frac{\partial c_{A,II}}{\partial t} &= F'_A \left(-v_{A,II} \frac{\partial c_{A,II}}{\partial z} + D \frac{\partial^2 c_{A,II}}{\partial z^2} \right) \\
\frac{\partial c_{B,II}}{\partial t} &= F'_B \left(-v_{B,II} \frac{\partial c_{B,II}}{\partial z} + D \frac{\partial^2 c_{B,II}}{\partial z^2} \right) \\
III: \quad z \in [2,3] \quad \frac{\partial c_{A,III}}{\partial t} &= F'_A \left(-v_{A,III} \frac{\partial c_{A,III}}{\partial z} + D \frac{\partial^2 c_{A,III}}{\partial z^2} \right) \\
\frac{\partial c_{B,III}}{\partial t} &= F'_B \left(-v_{B,III} \frac{\partial c_{B,III}}{\partial z} + D \frac{\partial^2 c_{B,III}}{\partial z^2} \right) \\
IV: \quad z \in [3,4] \quad \frac{\partial c_{A,IV}}{\partial t} &= F'_A \left(-v_{A,IV} \frac{\partial c_{A,IV}}{\partial z} + D \frac{\partial^2 c_{A,IV}}{\partial z^2} \right) \\
\frac{\partial c_{B,IV}}{\partial t} &= F'_B \left(-v_{B,IV} \frac{\partial c_{B,IV}}{\partial z} + D \frac{\partial^2 c_{B,IV}}{\partial z^2} \right) .
\end{aligned}
\tag{3.31}
$$

The initial condition of (3.31) is

$$
\begin{aligned}
c_{A,j}(0,z) &= c_{0,A,j}(z) \\
c_{B,j}(0,z) &= c_{0,B,j}(z) .
\end{aligned}
\tag{3.32}
$$

Two kinds of boundary conditions have to be considered for the closed–loop TMB process. First, because of the closed–loop counter–current of the adsorbent and the solvent, the concentration profiles $c_A(t, z)$ and $c_B(t, z)$ are continuous with respect to $z \in [0, 4]$.

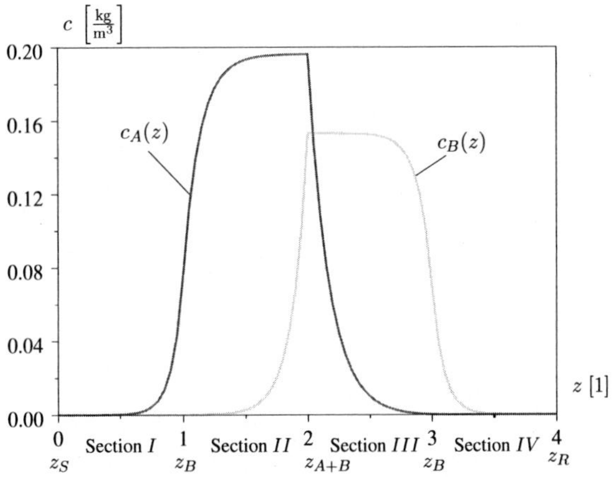

Figure 3.9: Concentration profiles of the TMB process with a closed–loop separation column

This yields the following boundary conditions:

$$\begin{aligned}
z = 0: \quad c_{A,I}(t,0) &= c_{A,IV}(t,4) \\
c_{B,I}(t,0) &= c_{B,IV}(t,4) \\
z = 1: \quad c_{A,I}(t,1) &= c_{A,II}(t,1) \\
c_{B,I}(t,1) &= c_{B,II}(t,1) \\
z = 2: \quad c_{A,II}(t,2) &= c_{A,III}(t,2) \\
c_{B,II}(t,2) &= c_{B,III}(t,2) \\
z = 3: \quad c_{A,III}(t,3) &= c_{A,IV}(t,3) \\
c_{B,III}(t,3) &= c_{B,IV}(t,3)\,.
\end{aligned} \tag{3.33}$$

Second, the mass flow balance at the inlet and outlet positions $z = 0, 1, 2, 3$ yields the following differential boundary conditions:

$$
\begin{aligned}
z=0: \quad & \left.\frac{\partial c_{A,I}}{\partial z}\right|_{z=0} - \left.\frac{\partial c_{A,IV}}{\partial z}\right|_{z=4} &=& \frac{v_{A,I}-v_{A,IV}}{D}\cdot c_{A,I}(t,0) \\
& \left.\frac{\partial c_{B,I}}{\partial z}\right|_{z=0} - \left.\frac{\partial c_{B,IV}}{\partial z}\right|_{z=4} &=& \frac{v_{B,I}-v_{B,IV}}{D}\cdot c_{B,I}(t,0) \\
z=1: \quad & \left.\frac{\partial c_{A,II}}{\partial z}\right|_{z=1} - \left.\frac{\partial c_{A,I}}{\partial z}\right|_{z=1} &=& 0 \\
& \left.\frac{\partial c_{B,II}}{\partial z}\right|_{z=1} - \left.\frac{\partial c_{B,I}}{\partial z}\right|_{z=1} &=& 0 \\
z=2: \quad & \left.\frac{\partial c_{A,III}}{\partial z}\right|_{z=2} - \left.\frac{\partial c_{A,II}}{\partial z}\right|_{z=2} &=& \frac{v_{A,III}-v_{A,II}}{D}\cdot \left(c_{A,III}(t,2)-c_{in,A}\right) \\
& \left.\frac{\partial c_{B,III}}{\partial z}\right|_{z=2} - \left.\frac{\partial c_{B,III}}{\partial z}\right|_{z=2} &=& \frac{v_{B,III}-v_{B,II}}{D}\cdot \left(c_{B,III}(t,2)-c_{in,B}\right) \\
z=3: \quad & \left.\frac{\partial c_{A,IV}}{\partial z}\right|_{z=3} - \left.\frac{\partial c_{A,III}}{\partial z}\right|_{z=3} &=& 0 \\
& \left.\frac{\partial c_{B,IV}}{\partial z}\right|_{z=3} - \left.\frac{\partial c_{B,III}}{\partial z}\right|_{z=3} &=& 0\,.
\end{aligned}
\tag{3.34}
$$

$c_{in,A}$ and $c_{in,B}$ are the concentrations of the components A and B in the feed inlet stream.

Together with the boundary conditions (3.33) and (3.34) the differential equations (3.31) form a cyclic system with respect to z. Therefore, the boundary conditions (3.33) and (3.34) are called cyclic boundary conditions. The complete TMB model is given by Equations (3.31), (3.32), (3.33) and (3.34).

3.3.2 Solution of the fluid dynamical model

The analytic solution of the TMB model (3.29) can be explicitly derived under the assumptions of linear adsorption. Then, the concentration profiles do not depend upon each other, i.e. $c_A(t,z) \neq f(c_B(t,z))$ and $c_B(t,z) \neq f(c_A(t,z))$.

The analytic solution of the TMB model (3.31) is composed of a dynamical and a stationary part:

$$
\begin{aligned}
c_{A,j}(t,z) &= c_{dyn,A,j}(t,z) + c_{stat,A,j}(z) \\
c_{B,j}(t,z) &= c_{dyn,B,j}(t,z) + c_{stat,B,j}(z)\,.
\end{aligned}
$$

The analytic expressions for the dynamical and the stationary part are derived based on the approach presented in Section 3.2.2. The resulting general solution of Equation (3.31) is

$$
\begin{aligned}
c_{A,j}(t,z) &= \sum_{i=1}^{\infty} \left(\mathrm{e}^{\lambda_{A,j,i}\, t} \left(p_{1,A,j,i} \cdot \mathrm{e}^{q_{1,A,j,i}\, z} + p_{2,A,j,i} \cdot \mathrm{e}^{q_{2,A,j,i}\, z} \right) \right) \\
&+ r_{A,j} \cdot \mathrm{e}^{\frac{v_{A,j}}{D} z} + s_{A,j} \\
c_{B,j}(t,z) &= \sum_{i=1}^{\infty} \left(\mathrm{e}^{\lambda_{B,j,i}\, t} \left(p_{1,B,j,i} \cdot \mathrm{e}^{q_{1,B,j,i}\, z} + p_{2,B,j,i} \cdot \mathrm{e}^{q_{2,B,j,i}\, z} \right) \right) \\
&+ r_{B,j} \cdot \mathrm{e}^{\frac{v_{B,j}}{D} z} + s_{B,j}\,,
\end{aligned}
\tag{3.35}
$$

for the sections $j = I, II, III, IV$.

The general solution (3.35) has to fulfil the initial conditions (3.32) for $t = 0$ for each TMB section. The explicit expression of the initial condition is given in Appendix C.1. Equation (C.1) shows that the parameters $p_{1,*,j,i}$, $q_{1,*,j,i}$, $p_{2,*,j,i}$ and $q_{2,*,j,i}$ and thereby the spectrum of $\lambda_{*,j,i}$, $i = 1, 2, ..., \infty$ of the general solution (3.35) are prescribed by the initial concentration profiles $c_{*,j}(0, z)$. The parameters $r_{*,j}$ and $s_{*,j}$, however, are determined by the constant operation parameters of the TMB process. The operation parameters are the relative flow velocities $v_{*,j}$ and the feed inlet concentrations $c_{*,in}$ ($* = A, B,\ j = I, II, III, IV$). The following Section 3.3.3 presents the derivation of the parameters $r_{*,j}$ and $s_{*,j}$.

3.3.3 Stationary model solution

The stationary solution of Equation (3.31) is given by

$$
c_{stat,*,j}(z) = r_{*,j} \cdot \mathrm{e}^{\frac{v_{*,j}}{D} z} + s_{*,j}\,.
$$

It is the aim to find an analytic expression for the parameters $r_{*,j}$ and $s_{*,j}$ in dependence upon the TMB process parameters. In other words, it is the aim to find the explicit functional expression of the stationary concentration profiles $c_{stat,A,j}(z)$ and $c_{stat,B,j}(z)$. It is assumed that the concentrations $c_A(t, z)$ and $c_B(t, z)$ do not depend upon each other and the general solution (3.35) as well as the boundary conditions (3.33) and (3.34), and the initial conditions (C.1) have the same form for both components A and B. Therefore, the expressions can be given without an explicit specification of the considered component $*$ and the hereafter derived stationary solution of the TMB process model applies to both components.

The stationary TMB model solution for the sections $j = I, II, III, IV$ and for the component $*$ is obtained from the stationary part of Equation (3.35):

$$\begin{aligned} I &: z \in [0,1], & c_{stat,*,I}(z) &= r_{*,I} \cdot e^{\frac{v_{*,I}}{D} z} + s_{*,I} \\ II &: z \in [1,2], & c_{stat,*,II}(z) &= r_{*,II} \cdot e^{\frac{v_{*,II}}{D} (z-1)} + s_{*,II} \\ III &: z \in [2,3], & c_{stat,*,III}(z) &= r_{*,III} \cdot e^{\frac{v_{*,III}}{D} (z-2)} + s_{*,III} \\ IV &: z \in [3,4], & c_{stat,*,IV}(z) &= r_{*,IV} \cdot e^{\frac{v_{*,IV}}{D} (z-3)} + s_{*,IV} . \end{aligned} \tag{3.36}$$

If the eight parameters $r_{*,j}$, $s_{*,j}$ are known, the solution is determined.

Equation (3.36) has to fulfil the boundary conditions (3.33) and (3.34). The application of Equation (3.36) to Equation (3.33) leads to the following four equations involving $r_{*,j}$ and $s_{*,j}$:

$$\begin{aligned} z = 0 :\ & r_{*,I} + s_{*,I} = r_{*,IV} \cdot e^{\frac{v_{*,IV}}{D}} + s_{*,IV} \\ z = 1 :\ & r_{*,II} + s_{*,II} = r_{*,I} \cdot e^{\frac{v_{*,I}}{D}} + s_{*,I} \\ z = 2 :\ & r_{*,III} + s_{*,III} = r_{*,II} \cdot e^{\frac{v_{*,II}}{D}} + s_{*,II} \\ z = 3 :\ & r_{*,IV} + s_{*,IV} = r_{*,III} \cdot e^{\frac{v_{*,III}}{D}} + s_{*,III} . \end{aligned} \tag{3.37}$$

With Equation (3.34) the following four equations are obtained:

$$\begin{aligned} z = 0 :\ & r_{*,I} \cdot \frac{v_{*,I}}{D} - r_{*,IV} \cdot \frac{v_{*,IV}}{D} \cdot e^{\frac{v_{*,IV}}{D}} = \frac{v_{*,I} - v_{*,IV}}{D} \cdot (r_{*,I} + s_{*,I}) \\ z = 1 :\ & r_{*,I} \cdot \frac{v_{*,I}}{D} \cdot e^{\frac{v_{*,I}}{D}} = r_{*,II} \cdot \frac{v_{*,II}}{D} \\ z = 2 :\ & r_{*,III} \cdot \frac{v_{*,III}}{D} - r_{*,II} \cdot \frac{v_{*,II}}{D} \cdot e^{\frac{v_{*,II}}{D}} = \frac{v_{*,III} - v_{*,II}}{D} \cdot (r_{*,III} + s_{*,III} - c_{*,in}) \\ z = 3 :\ & r_{*,III} \cdot \frac{v_{*,III}}{D} \cdot e^{\frac{v_{*,III}}{D}} = r_{*,IV} \cdot \frac{v_{*,IV}}{D} . \end{aligned} \tag{3.38}$$

The system of Equations (3.37) and (3.38) can be solved analytically. The explicit expressions for the parameters $r_{*,j}$ and $s_{*,j}$ are given in Appendix C.1. Figure 3.9 shows the two stationary TMB concentration profiles $c_{stat,A,j}(z)$ and $c_{stat,B,j}(z)$ of an example, which is given in the Appendix C.2.

3.4 Continuous dynamics

Both the SMB and the VARICOL are chromatographic separation processes with a simulated counterflow. The processes are characterised by the superposition of a continuous–variable mass transport and mass exchange, and the discrete–event switching of the inlet and outlet ports on the circle of separation columns. Hence, for the physical modelling of the system, the two system classes, i.e. the continuous mass transfer and the discrete port switching, have to be considered separately before the whole system can be modelled. In this section, the physical model of the mass transfer in the circle of separation columns is derived, assuming fixed port positions. The resulting model can be applied to both the SMB and the VARICOL.

The fluid dynamical model describes the continuous part of SCC processes. It is derived based on the single–column model and the node balances of the column interconnections. The resulting model consists of n_c partial differential equations for each component $*$, which are coupled via the node balances. The parameters of this system depend upon the internal fluid flow rates, the positions of the inlet and outlet ports with respect to the separation columns, and the feed inlet concentrations.

Physical model of a single separation column. The continuous mass transfer in an SMB or VARICOL process takes place in a circle of interconnected separation columns. For each column $m = 1, 2, \ldots, n_c$ the Equilibrium–Dispersive–Model (3.13) is used to describe the concentration distribution $c_{*,m} = c_{*,m}(t, z)$ for the component $* = A, B$:

$$\frac{\partial c_{*,m}}{\partial t} = F'_{*,m} \left(-u_m \frac{\partial c_{*,m}}{\partial z} + D \frac{\partial^2 c_{*,m}}{\partial z^2} \right) \text{, for } z \in \left[z_{in,m}, z_{out,m} \right]. \tag{3.39}$$

The index m specifies the considered column $1, 2, 3, \ldots, n_c$. $F'_{*,m}$ describes the influence of the adsorption:

$$F'_{*,m} = \frac{1}{1+F \frac{\partial f_{Eq,*}}{\partial c_{*,m}}} .$$

The void fraction of the fixed bed with $F_m = \frac{1-\varepsilon_m}{\varepsilon_m}$ is determined by the package porosity ε_m in the column m. $\frac{\partial f_{Eq,*}}{\partial c_{*,m}}$ is the slope of the adsorption isotherm, which takes constant values for a linear adsorption. u_m is the flow velocity in the column m

$$u_m = \frac{\dot{m}_m}{\rho_m \, A_m}\,, \tag{3.40}$$

where $\dot{m}_m$ is the solvent mass flow in the SCC section, in which the column is actually sited, ρ_m is the solvent density and A_m is the cross section area of the porous fixed bed in the column m. D is the diffusion coefficient, which is considered to be time–invariant and which takes the same value for all separation columns and components. Equation (3.39) is an implicit description of the concentration profile $c_{*,m}(t,z)$ in the column m for the component $* = A, B$ as a function of time t and the spatial range $z \in [z_{in,m}, z_{out,m}]$. $z_{in,m}$ denotes the fluid inlet and $z_{out,m}$ the fluid outlet position of the considered column.

For a complete description of the continuous mass transfer, the initial conditions have to be given, which in the case of an empty SCC plant are

$$c_{*,m}(0,z) = 0 \;\text{ for } * = A, B\,, \;\; m \in \{1, 2, \ldots, n_c\}\,.$$

Applying the moving spatial coordinate system representation of $c_{*,m}(t,z)$, it has to be considered that the initial conditions perform a jump with respect to z in the moment of port switching (Kleinert and Lunze, 2002). This phenomenon will be discussed in more detail in Sections 3.6.2 and 3.6.3.

The boundary conditions (3.17a) and (3.17b) which were derived for the single–column separation model, are applied here:

$$\boxed{\begin{aligned} z = z_{in,m}: &\quad c_{*,in,m}(t) = \left(c_{*,m} - \tfrac{D}{u_m}\tfrac{\partial c_{*,m}}{\partial z}\right)\Big|_{z_{in,m}} \\ z = z_{out,m}: &\quad c_{*,out,m}(t) = \left(c_{*,m} - \tfrac{D}{u_m}\tfrac{\partial c_{*,m}}{\partial z}\right)\Big|_{z_{out,m}} \,. \end{aligned}} \tag{3.41}$$

Node balances for the column interconnections. To describe the physical behaviour of the complete SCC process the mathematical systems consisting of the differential equation (3.39) and the boundary conditions (3.41) are coupled by the node balances of the column interconnections. If non–zero extra–column volumes $V_{d,m}$ of the column interconnections are considered, the outlet concentration of the upstream column m reaches the inlet of the downstream column $(m+1)_{\text{modulo } nc}$ with a time delay of $T_{d,m}$, which depends upon $V_{d,m}$ and the fluid flow rate in the considered interconnection. The node balance of the interconnection of two columns $[m, m+1]_{\text{modulo } n_c}$, to which actually *no* inlet or outlet port is connected, yields

$$c_{*,in,m+1}(t) = c_{*,out,m}(t - T_{d,m})\,. \tag{3.42}$$

The internal mass flow rates underlie a dynamical change in case of a closed–loop control of the operation point. This means that the resulting time delays $T_{d,m}$ are time–varying in dependence upon the internal fluid flow rates of the section, in which the considered interconnection is actually sited.

If a *product outlet port* is attached to the interconnection of the columns labled with m and $(m+1)_{\text{modulo } n_c}$, i.e. at the positions z_A or z_B, the result of the node balance corresponds to that of Equation (3.42):

$$\begin{aligned} c_{*,in,II}(t) &= c_{*,out,I}(t - T_{d,z_A}) \\ c_{*,in,IV}(t) &= c_{*,out,III}(t - T_{d,z_B})\,. \end{aligned} \tag{3.43}$$

$c_{*,in,j}$ and $c_{*,out,j}$ specify the concentrations of the fluid leaving or entering the specified section j, respectively. T_{d,z_A} and T_{d,z_B} specify the actual time delay which is associated with the considered column interconnection.

At the inlet positions z_S and z_{A+B}, which are mixing points of different fluids, the node balances yield

$$c_{in,I,*}(t) = \frac{\dot{m}_{IV}}{\dot{m}_I}\, c_{out,IV,*}(t - T_{d,z_S})\,, \tag{3.44}$$

for $z = z_S$ and

$$c_{in,III,*}(t) = \frac{\dot{m}_{II}}{\dot{m}_{III}}\, c_{out,II,*}(t - T_{d,z_{A+B}}) + \frac{\dot{m}_{A+B}}{\dot{m}_{III}}\, c_{A+B,*}(t)\,, \tag{3.45}$$

for $z = z_{A+B}$. $c_{A+B,*}(t)$ specifies the concentration of the component $* = A, B$ in the feed inlet stream.

Fluid dynamical model for SCC processes. The fluid dynamical model for the circle of separation columns can be represented using the physical model for a single separation column and

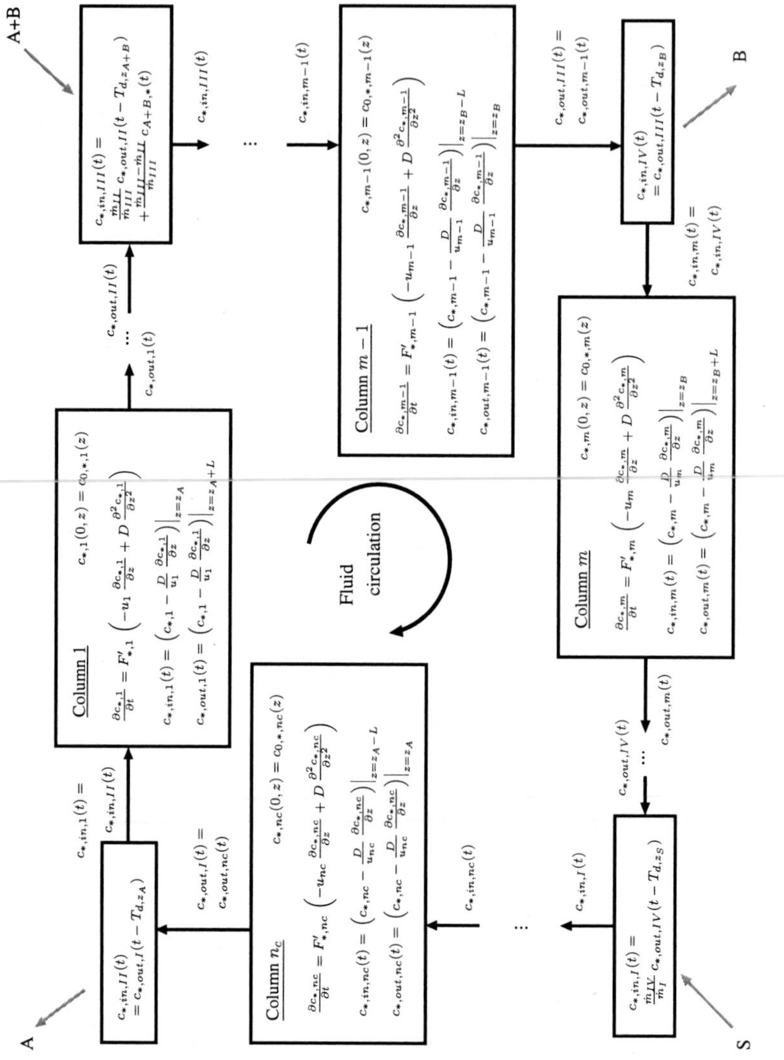

Figure 3.10: Physical model of the continuous–variable SCC subsystem considering a closed recycling loop

the node balances for the column interconnections. In the following, this model is referred to as the physical model of the continuous dynamics of SCC processes. Figure 3.10 shows the block diagram of the process model considering a closed recycling loop. The positions of the inlet and outlet ports on the corners of the loop are indicated by A, $A+B$, B and S. These positions are chosen arbitrarily and change with each port switching. The block diagram shows, how the partial differential equations describing the concentrations $c_{*,m}(t,z)$ of the column m are connected via the node balances. The impact of the continuous control input, i.e. the internal flow velocities u_m, onto the system becomes visible from the model equations in Figure 3.10.

To describe the impact of the port positions z_S, z_A, z_{A+B} and z_B onto the model, the discrete variable

$$\mathsf{w} = \begin{pmatrix} z_S \\ z_A \\ z_{A+B} \\ z_B \end{pmatrix}, \tag{3.46}$$

with

$$\begin{aligned} z_S &\in \{0,1,\ldots,n_c-1\}, \\ z_A, z_{A+B}, z_B &\in \{1,\ldots,n_c\} \end{aligned} \tag{3.47}$$

is introduced. w indicates the port positions z_S, z_A, z_{A+B} and z_B with respect to each column $m = 1, 2, \ldots, n_c$. The position z_S of the solvent injection is determined by the spatial coordinate, which refers to the column *inlet* position $z = m-1$ of the separation column m. The remaining port positions z_A, z_{A+B} and z_B are determined by the spatial coordinate describing the *outlet* position $z = m$ of the column m.

Because the port positions are changed with respect to the column positions when a port switching signal occurs, the port positions and, hence, the variable w, are a function of time. With each switching of one port the internal fluid flow in one separation column changes. Hence, the dynamical behaviour of the continuous subsystem changes. Considering the representation in Figure 3.10, it becomes clear that for each possible combination of port positions on the circle of columns, one unique model of the continuous dynamics is obtained.

To describe the impact of the port positions onto the continuous dynamics the discrete variable w has to be converted to a continuous signal. Because the port positions together with the internal fluid flow rates determine the flow velocity in each column $m = 1, 2, \ldots, n_c$, the variable w is converted to the vector

$$w(t) = \begin{pmatrix} u_1(t) \\ \vdots \\ u_m(t) \\ \vdots \\ u_{n_c}(t) \end{pmatrix},$$

which assigns to each column m the flow velocity $u_m(t)$ according to w.

A switching of the port positions changes the parameters of the continuous subsystem by a discrete action, which leads to an instantaneous change of the continuous dynamics. Considering the moving spatial coordinate representation of the concentration profiles, the switching of the solvent inlet port S leads to a discrete jump of the continuous state.

Remark. The continuous SCC subsystem can be modelled using a spatial coordinate z of which the origin is tied to a specific column (fixed spatial coordinate) or to the position of a port, e.g. the solvent inlet (moving spatial coordinate). It is reasonable to choose a fixed spatial coordinate if the column characteristics and the extra–column volumes $V_{d,m}$ show a significant variance. Then, the assignment of the model parameters, which depend upon the column characteristics and $V_{d,m}$, with respect to z remains unchanged if a port switching takes place. If the variance of the column characteristics and $V_{d,m}$ can be neglected, the corresponding parameters are constant for $z \in [0, n_c L]$ for both the application of the fixed and the moving spatial coordinate.

Compact representation of the continuous dynamics. The component distribution in each column is represented using the vector

$$\mathbf{c}_m(t,z) = \begin{pmatrix} c_{A,m}(t,z) \\ c_{B,m}(t,z) \end{pmatrix}.$$

Based on this representation, the state vector $\boldsymbol{x}$ of the continuous dynamics of SCC processes is introduced:

$$\boldsymbol{x}(t,z) = \begin{pmatrix} \mathbf{c}_1(t,z) \\ \vdots \\ \mathbf{c}_m(t,z) \\ \vdots \\ \mathbf{c}_{n_c}(t,z) \end{pmatrix}. \tag{3.48}$$

The introduction of the partial temporal derivative of the state vector

$$\dot{\boldsymbol{x}} := \frac{\partial \boldsymbol{x}(t,z)}{\partial t} \tag{3.49}$$

allows for the representation of the continuous dynamics, which is given by Equation (3.39), the boundary conditions (3.41) and the node balances (3.43) to (3.45), by a single vector operator $\boldsymbol{f}(\boldsymbol{x}(t,z), \boldsymbol{u}(t), \boldsymbol{d}(t), w(t))$. This is possible, because the model equations of the columns, the boundary conditions and the node balances form a cyclic system, which can be lumped to a system of equations. The variables of the operator are the state vector $\boldsymbol{x}(t,z)$, the internal fluid flow rates described by the control input vector $\boldsymbol{u}(t)$, the assignment vector $w(t)$, and the feed inlet concentrations and the adsorbent porosity, represented by the disturbance vector $\boldsymbol{d}(t)$. The flow velocities in each separation column are determined by the vector of the internal fluid flow rates

$$\boldsymbol{u}(t) = \begin{pmatrix} \dot{m}_I(t) \\ \dot{m}_{II}(t) \\ \dot{m}_{III}(t) \\ \dot{m}_{IV}(t) \end{pmatrix}, \tag{3.50}$$

which can be transformed to the flow velocities $w(t)$ applying Equation (3.40) and considering the port positions with respect to the columns. The disturbance signals are lumped in the vector

$$\boldsymbol{d}(t) = \begin{pmatrix} c_{A+B,A}(t) \\ c_{A+B,B}(t) \\ \varepsilon(t) \end{pmatrix}. \tag{3.51}$$

Hence, the state equation of the continuous dynamics can be given by

$$\dot{\boldsymbol{x}} = \boldsymbol{f}(\boldsymbol{x}(t,z), \boldsymbol{u}(t), \boldsymbol{d}(t), w(t))\,. \tag{3.52}$$

Considering the concentrations at the column outlet positions as the output variable, the output equation of the continuous dynamics is

$$\boldsymbol{y}(t) = \begin{pmatrix} y_1(t) \\ \vdots \\ y_m(t) \\ \vdots \\ y_{n_c}(t) \end{pmatrix} = \begin{pmatrix} c_1(t,L) \\ \vdots \\ c_m(t,L) \\ \vdots \\ c_{n_c}(t,L) \end{pmatrix} .$$

Figure 3.11 shows a block diagram of the compact continuous dynamics representation.

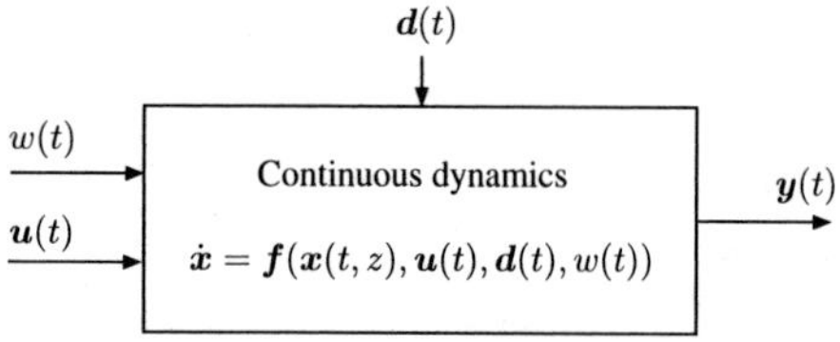

Figure 3.11: Compact representation of the continuous SCC subsystem

Remark. The presented model is a physical model of the continuous mass transport in an SCC separation process. The model is considered as the continuous subsystem of the process. The presented model is very complex. To the knowledge of the author, no analytical solution to this model is published in literature up to now. The presented modelling is a framework for the physical modelling of the continuous dynamics of SCC processes. The framework is commonly used to set up a numerical simulation of the processes (Giese, 2002; Kleinert, 2002). It allows for the application of various models for the chromatographic separation in single separation columns, of which here the Equilibrium–Dispersive–Model is considered, and of various boundary conditions for the column inlet and outlet positions.

If a significant variance of the column characteristics due to different package porosities as well as different extra–column volumes have to be considered, the cyclic behaviour of SCC processes with time–invariant switching patterns does not occur with respect to a switching period but with respect to a switching cycle. In practical applications, the variance is low and, therefore can be neglected. This leads to the following assumption for all SCC processes considered within this thesis:

Assumption 3.4.1 *The difference of the package porosity in the SCC columns and of the extra–column volumes of the column interconnections of an SCC plant is low. Therefore, it is assumed that all SCC columns have the same mass transport characteristics and all extra–column volumes are the same.* □

3.5 Discrete dynamics

The switching of the inlet and outlet ports of an SCC process is a discrete–event system. The input signal consists of the discrete port switching signals ("switch port $*$" or "do not switch port $*$") and the output signal of the discrete positions of the inlet and outlet ports (Figure 3.12).

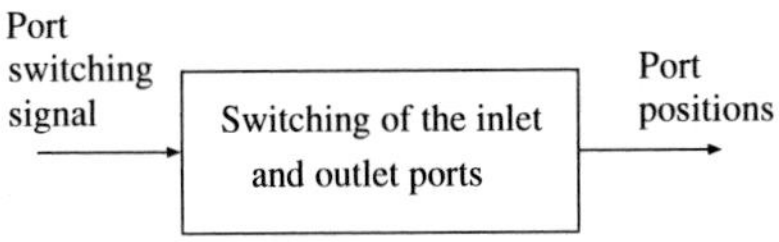

Figure 3.12: Port positioning and port switching of an SCC process

The modelling of the dynamical behaviour of the port switching is presented in Section 3.5.1. A model according to Equation (3.3) is introduced for the most general SCC process, which is a VARICOL process with n_c separation columns and a time–varying switching pattern, to which the fixed spatial coordinate system representation is applied. The discrete model is derived regarding Assumption 2.4.1. In Section 3.5.2 the result is applied to the modelling of the discrete subsystems of SMB processes for a fixed and a moving spatial coordinate. In Section 3.5.3 the model is applied to VARICOL processes considering time–varying and time–invariant switching patterns for both fixed and moving spatial coordinate systems.

3.5.1 Discrete–event subsystem

The discrete–event subsystem of SCC processes describes the switching of the inlet and outlet ports on the circle of separation columns in dependence upon an external port switching command. To analyse the system properties and to derive a model of the dynamical behaviour, a four column SCC plant is considered as an example. Figure 3.13 shows the circle of the columns $m = 1, 2, 3, 4$ and one possible combination of port positions z_S, z_A, z_{A+B} and z_B.

The number of columns $n_{c,j} \in \mathbb{N}^0$, $j = I, II, III, IV$ in the SCC sections is referred to as the section length according to the concept introduced in Section 2.4.1. It is determined by the port positions. For the example given in Figure 3.13 the section length is shown in Table 3.2.

Based on Assumption 2.4.1 the following conditions are obtained for the section length:

$$n_{c,j} \overset{!}{\geq} 0 \text{ for } j = I, II, IV \,, \tag{3.53}$$

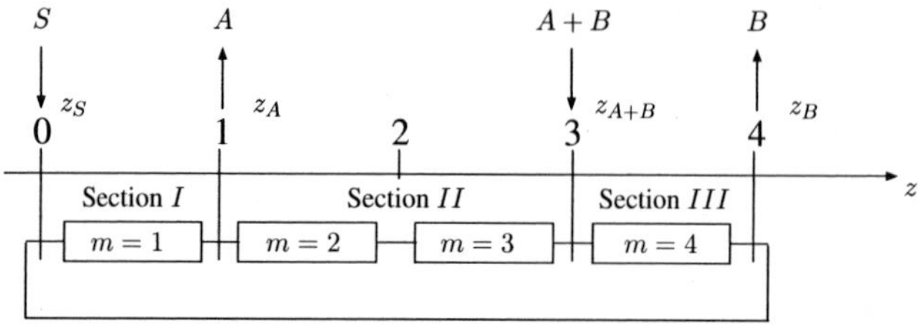

Figure 3.13: Positions of the inlet and outlet ports on the circle of separation columns

Section	Port positions	Section length
I	$z_S = 0$, $z_A = 1$	$n_{c,I} = 1$
II	$z_A = 1$, $z_{A+B} = 3$	$n_{c,II} = 2$
III	$z_{A+B} = 3$, $z_B = 4$	$n_{c,III} = 1$
IV	$z_B = 4$, $z_S = 0$	$n_{c,IV} = 0$

Table 3.2: Number of columns per section

$$n_{c,III} \overset{!}{\geq} 1\,. \tag{3.54}$$

To analyse the dynamical behaviour of the port switching signal it is now assumed that the ports S, A and B are switched by one column length. Then, the new port positions shown in Figure 3.14 are obtained. The new port positions and the section lengths are indicated in Table 3.3.

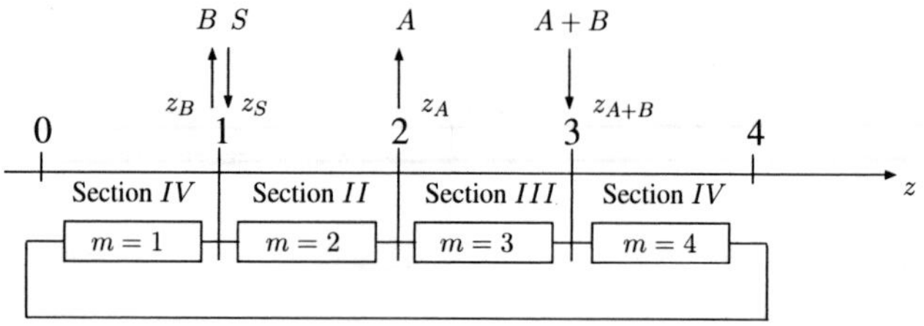

Figure 3.14: Port positions after the switching of S, A and B with respect to Figure 3.13

To describe the transition from the configuration shown in Figure 3.13 to that shown in Figure 3.14 it is convenient to introduce the relative port positions $z_{rel,A}$, $z_{rel,A+B}$, and $z_{rel,B}$. The relative port positions indicate the number of columns between the solvent inlet S and the respective port A, $A + B$ or B. In terms of the section length $n_{c,j}$, $j = I, II, III, IV$ the relative port positions are determined by

Section	Port positions	Section length
I	$z_S = 1$, $z_A = 2$	$n_{c,I} = 1$
II	$z_A = 2$, $z_{A+B} = 3$	$n_{c,II} = 1$
III	$z_{A+B} = 3$, $z_B = 1$	$n_{c,III} = 2$
IV	$z_B = 1$, $z_S = 1$	$n_{c,IV} = 0$

Table 3.3: Number of columns per section

$$\begin{aligned} z_{rel,A} &:= n_{c,I} \\ z_{rel,A+B} &:= n_{c,I} + n_{c,II} \\ z_{rel,B} &:= n_{c,I} + n_{c,II} + n_{c,III} \,. \end{aligned} \tag{3.55}$$

Because a section length $n_{c,j} \in \mathbb{N}_0$, $j = I, II, III, IV$ is an integer, the relative port positions are also integer values:

$$z_{rel,*} \in \{0, 1, \ldots, n_c\}, \quad * = A, A+B, B \,.$$

With respect to Assumption 2.4.1 the relative port positions $z_{rel,A}$, $z_{rel,A+B}$ and $z_{rel,B}$ have to fulfil the following inequality:

$$0 \leq z_{rel,A} \leq z_{rel,A+B} < z_{rel,B} \,. \tag{3.56}$$

Modelling task. It is the aim to find a formal description of the port position transitions due to a switching command, using the representation of discrete–event systems proposed in Section 3.1.1. The following **modelling task** is considered:

Task 3.5.1 *Find suitable input, state and output variables* v, z *and* w *and the corresponding state transition and output relation* G : $G(\mathsf{z}, \mathsf{v}) \mapsto \mathsf{z}'$ *and* H : $H(\mathsf{z}, \mathsf{v}) \mapsto \mathsf{w}$*, which describe the dynamical behaviour of the port switching of SCC processes for which Assumption 2.4.1 holds.*
□

Solution of the modelling task. To describe the dynamics of the discrete subsystem the discrete counter $k \in \mathbb{N}$ of the transitions is introduced.

According to the variable w, which was introduced in Section 3.4 in Equation (3.46), the coordinates of the port positions define the output variable w of the discrete subsystem

$$\mathsf{w}(k) = \begin{pmatrix} z_S(k) \\ z_A(k) \\ z_{A+B}(k) \\ z_B(k) \end{pmatrix}, \tag{3.57}$$

whose elements $\mathsf{w}_i \in \mathbb{N}_0$, $i = 1, 2, 3, 4$ take integer values. The input v is the vector

$$\mathsf{v}(k) = \begin{pmatrix} E_S(k) \\ E_A(k) \\ E_{A+B}(k) \\ E_B(k) \end{pmatrix}, \tag{3.58}$$

of which the elements

$$E_*(k) \in \{0, 1\}\ , \quad * = S, A, A+B, B$$

are the switching signals for the solvent inlet S, the product outlet A or B or the feed inlet $A+B$, respectively. If $E_* = 1$ the port $*$ is switched while regarding Assumption 2.4.1, else if $E_* = 0$ the port remains at its position. If any $E_* = 1$ occurs and a new port switching event is triggered, the counter k is raised by 1.

For the solution of Task 3.5.1 the discrete state variable

$$\mathsf{z}(k) = \begin{pmatrix} z_S(k) \\ z_S(k) + z_{rel,A}(k) \\ z_S(k) + z_{rel,A+B}(k) \\ z_S(k) + z_{rel,B}(k) \end{pmatrix} \tag{3.59}$$

is introduced. $z_{rel,A}$, $z_{rel,A+B}$ and $z_{rel,B}$ specify the relative port position as defined by Equation (3.55). Because the relative port positions and the port position z_S are integer values, the elements z_i, $i = 1, 2, 3, 4$ are also integer values. Considering Equation (3.47) the following range for z_i is obtained:

$$\begin{aligned} \mathsf{z}_1 &\in \{0, 1, \ldots, n_c - 1\} \\ \mathsf{z}_i &\in \{1, \ldots, n_c\}\ , \quad i = 2, 3, 4\,. \end{aligned}$$

Example. The example shown in Figure 3.13 is described by the following state and output variable for $k = 1$:

$$\mathsf{z}(k=1) = \begin{pmatrix} 0 \\ 0+1 \\ 0+3 \\ 0+4 \end{pmatrix} = \begin{pmatrix} 0 \\ 1 \\ 3 \\ 4 \end{pmatrix}, \quad \mathsf{w}(k=1) = \begin{pmatrix} 0 \\ 1 \\ 3 \\ 4 \end{pmatrix}.$$

The following switching signal was assumed for $k = 1$:

$$\mathsf{v}(k=1) = \begin{pmatrix} 1 \\ 1 \\ 0 \\ 1 \end{pmatrix}.$$

The successor state $\mathsf{z}'(k = 1)$ is

$$\mathsf{z}(k=2) = \mathsf{z}'(k=1) = \begin{pmatrix} 1 \\ 1+1 \\ 1+2 \\ 1+4 \end{pmatrix} = \begin{pmatrix} 1 \\ 2 \\ 3 \\ 5 \end{pmatrix}.$$

Figure 3.14 shows the new port positions for $k = 2$. The new output variable is

$$\mathsf{w}(k=2) = \begin{pmatrix} 1 \\ 2 \\ 3 \\ 1 \end{pmatrix}.$$

Because the elements of v, z, z' and w are integer values, it is possible to apply arithmetic operations to the variables. Obviously, the difference of the successor state z' and the state z yields the input variable v:

$$\mathsf{z}'(k=1) - \mathsf{z}(k=1) = \mathsf{v}(k=1).$$

This allows to determine the successor state z' from the actual state z and input v:

$$\mathsf{z}'(k=1) = \mathsf{z}(k=1) + \mathsf{v}(k=1). \tag{3.60}$$

The transition of the discrete state can be represented by an automaton graph as is shown in Sections 3.5.2 and 3.5.3.

The output variable w is derived from the state variable as follows:

$$\begin{aligned} && \mathsf{w}(k) &= \mathsf{z}(k) \\ \text{if} \quad \mathsf{w}_i(k) \geq n_c\,, & \text{then} & \mathsf{w}_i(k) &= \mathsf{w}_i(k) - n_c\,, \quad i = 1,2,3 \\ \text{if} \quad \mathsf{w}_4(k) > n_c\,, & \text{then} & \mathsf{w}_4(k) &= \mathsf{w}_4(k) - n_c\,, \end{aligned} \tag{3.61}$$

where w_i, $i = 1,2,3,4$ specifies the elements of w. The range of w_i is determined by Equation (3.47).

State relation. The previously discussed example shows that the successor state z' can be determined from the state z and the input v by a simple addition. Hence, the state transition relation G according to Equation (3.3) can easily be derived from Equation (3.60). However, the restrictions to the port positions as specified by Equation (3.56) have to be regarded. Therefore, the conditionally executed actions $A1$, $A2$, $A3$ and $A4$ have to be formulated in addition to Equation (3.60) to obtain G:

$$\begin{aligned} G: & & & \mathsf{z}' = \mathsf{z} + \mathsf{v} \\ A1: & \text{ if } \mathsf{z}'_1 > \mathsf{z}'_2 & \text{then set } \mathsf{z}'_2 \text{ to } & \mathsf{z}'_2 = \mathsf{z}'_1 \\ & \text{ if } \mathsf{z}'_2 > \mathsf{z}'_3 & \text{then set } \mathsf{z}'_3 \text{ to } & \mathsf{z}'_3 = \mathsf{z}'_2 \\ A2: & \text{ if } \mathsf{z}'_3 \geq \mathsf{z}'_4 & \text{then set } \mathsf{z}'_4 \text{ to } & \mathsf{z}'_4 = \mathsf{z}'_3 + 1 \\ A3: & \text{ if } \mathsf{z}'_4 - \mathsf{z}'_1 \geq n_c & \text{then set } \mathsf{z}'_1 \text{ to } & \mathsf{z}'_1 = \mathsf{z}'_1 + 1 \\ & & \text{and } & \text{repeat } A1 \text{ and } A2 \\ A4: & \text{ if } \mathsf{z}'_1 \geq n_c & \text{then set } \mathsf{z}' \text{ to } & \mathsf{z}' = \mathsf{z}' - n_c\,, \end{aligned} \tag{3.62}$$

where z_i, z'_i, $i = 1,2,3,4$ specify the elements of z or z', respectively. To determine the successor state z', the conditions for the actions $A1$ to $A4$ have to be checked and the actions have to be executed in the given order. The actions have the following effect:

- ◇ $A1$ guarantees that none of the ports S, A and $A + B$ overtakes the next: If A is directly sited next to S and only S shall be switched, then A is also switched. The same holds for the pair $(A, A + B)$.

- ◇ $A2$ guarantees that there is at least one column in section III: If there is one column between $A + B$ and B and only $A + B$ shall be switched, B is also switched.

- ◇ $A3$ guarantees that if B is directly sited next to S and only B shall be switched, then S is also switched. Because this latter action can also cause a violation of the port order, the conditions for $A1$ and $A2$ have to be checked again.

- ◇ $A4$ guarantees that all state elements are reset if the number of z_1 reaches the number of columns n_c.

The proposed transition relation applies to SCC processes with $n_c \geq 4$ columns. The initial condition of z has to fulfil the restrictions (3.56) such that for $\mathsf{z}(0)$

$$\boxed{\mathsf{z}_1(0) \leq \mathsf{z}_2(0) \leq \mathsf{z}_3(0) < \mathsf{z}_4(0) \leq \mathsf{z}_1(0) + n_c} \tag{3.63}$$

holds.

Remark. Due to the actions $A1$, $A2$ and $A3$ it is possible that the input signal gives the switching command for one port only, but several ports are switched at a time, i.e. the system performs more than one transition.

Output relation. The output relation H is directly obtained from Equation (3.61):

$$\boxed{\begin{array}{llcl} H: & & \mathsf{w} & = \mathsf{z} \\ & \text{if } \mathsf{w}_i \geq n_c\,, \text{ then} & \mathsf{w}_i & = \mathsf{w}_i - n_c \quad i = 1,2,3 \\ & \text{if } \mathsf{w}_4 > n_c\,, \text{ then} & \mathsf{w}_4 & = \mathsf{w}_4 - n_c\,. \end{array}} \tag{3.64}$$

H applies to SCC processes with $n_c \geq 4$ columns.

The input variable v, the state variable z, the successor state z' and the output variable w together with the state transition relation G, the initial condition $\mathsf{z}(0)$ and the output relation H completely describe the dynamical behaviour of the port switching of an SCC process, which has n_c separation columns. The restrictions (3.53) and (3.54) to the port positions, which insure the functionality of the SCC process with respect to Assumption 2.4.1, are fulfilled if the condition (3.63) for the initial condition $\mathsf{z}(0)$ is met.

Solution 3.5.1 *The solution of Task 3.5.1 based on the choice of the input* v, *the state* z, *the successor state* z' *and the output* w, *by the Equations (3.58), (3.59) and (3.57), is given by the state transition relation* G *by Equation (3.62) and the output relation* H *by Equation (3.64), if the condition (3.63) for* $\mathsf{z}(0)$ *is fulfilled. Hence, the discrete–event model of the discrete SCC subsystem as shown in Figure 3.15 is found.* □

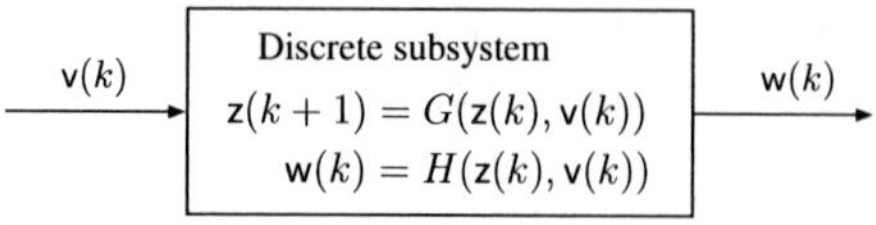

Figure 3.15: Discrete SCC subsystem

3.5.2 Discrete–event subsystem of the SMB process

Based on the discrete–event model derived in the previous section the discrete subsystems for SMB processes can be analysed. Because all ports of SMB processes are switched synchronously, the input signal v takes two values:

$$\text{port switching: } \mathsf{v} = \begin{pmatrix} 1 \\ 1 \\ 1 \\ 1 \end{pmatrix}, \quad \text{no port switching: } \mathsf{v} = \begin{pmatrix} 0 \\ 0 \\ 0 \\ 0 \end{pmatrix}.$$

Because no asynchronous port switching takes place it is sufficient to count the transitions of the discrete subsystem, in which all ports are switched synchronously at the same time, by k. Then, k corresponds to the counter of switching periods introduced in Section 2.3.2. In the following, the discrete SMB subsystem is analysed separately for a fixed and a moving spatial coordinate z.

Fixed spatial coordinate z. Because of the synchronous switching the section length $n_{c,j}$, $j = I, II, III, IV$ of SMB processes does not change. Therefore, the relative port positions $z_{rel,A}$, $z_{rel,A+B}$ and $z_{rel,B}$ are time–invariant. This means that for a given initial column setup $(n_{c,0,I}/n_{c,0,II}/n_{c,0,III}/n_{c,0,IV})$ the port position z_S is the only variable of the discrete state z:

$$\mathsf{z}(k) = \begin{pmatrix} z_S(k) \\ z_S(k) + z_{rel,A} \\ z_S(k) + z_{rel,A+B} \\ z_S(k) + z_{rel,B} \end{pmatrix}.$$

The transition relation G for this case is obtained from Equation (3.62). Because of the synchronous switching the actions $A1$ to $A3$ are omitted. The discrete subsystem of the SMB is a cyclic process and, therefore, the action $A4$ has to be taken into account:

$$\begin{aligned} G: & \qquad \mathsf{z}' = \mathsf{z} + \mathsf{v} \\ A4: \ \text{if} \ \ \mathsf{z}'_1 \geq n_c \ \ \text{then set } \mathsf{z}' \text{ to} & \ \ \mathsf{z}' = \mathsf{z}' - n_c \,, \end{aligned} \tag{3.65}$$

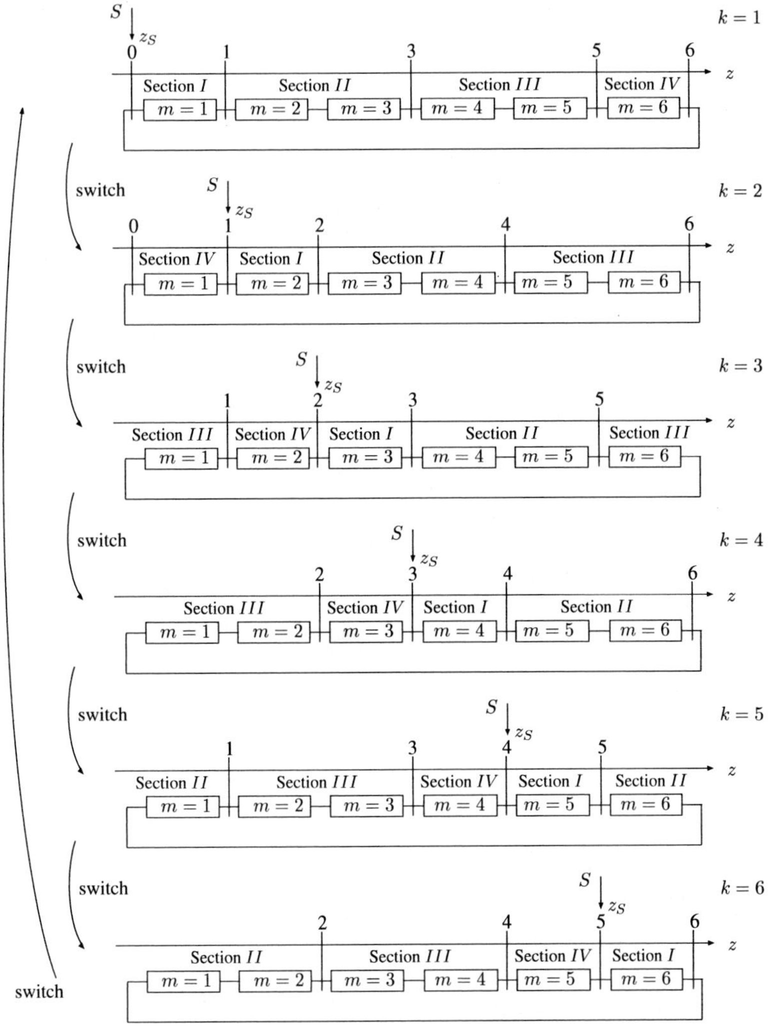

Figure 3.16: Sequence of port positions of a six column SMB plant (fixed spatial coordinate)

for $\mathsf{z}(0) = \mathsf{z}_0$. The output relation (3.64) remains unchanged.

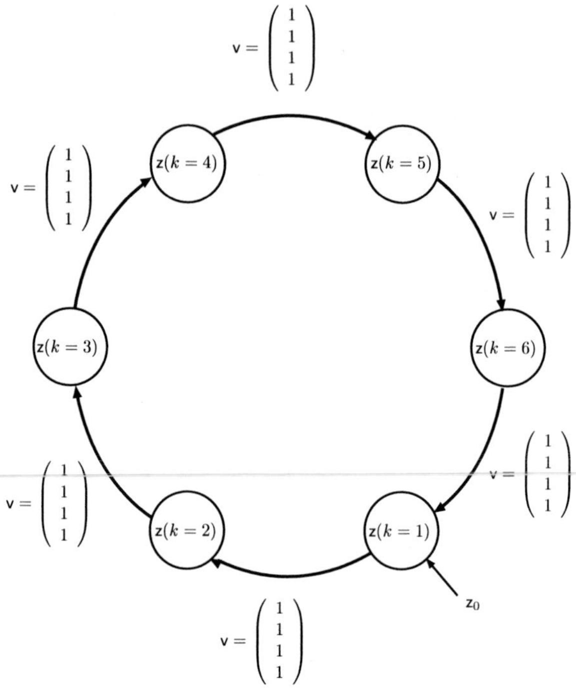

Figure 3.17: Automaton graph of the discrete subsystem of a six column SMB process using a fixed spatial coordinate system z

Figure 3.16 shows an example of an SMB process with $n_c = 6$ columns. A fixed coordinate system is used, which origin is placed to the inlet of the column $m = 1$. The figure shows that there are $n_c = 6$ possibilities to position the port S. Hence, the number of discrete states, which equals the number of discrete outputs, is equal to the number of columns n_c:

$$n_{\mathsf{z}} = n_{\mathsf{w}} = n_c \,.$$

The sequence of port positions shown in Figure 3.16 is represented by an automaton graph in Figure 3.17. The nodes of the graph represent the discrete states, and the edges represent the state transitions in case of port switching. The graph shows the cyclic behaviour of the discrete SMB subsystem in case of a fixed spatial coordinate system z.

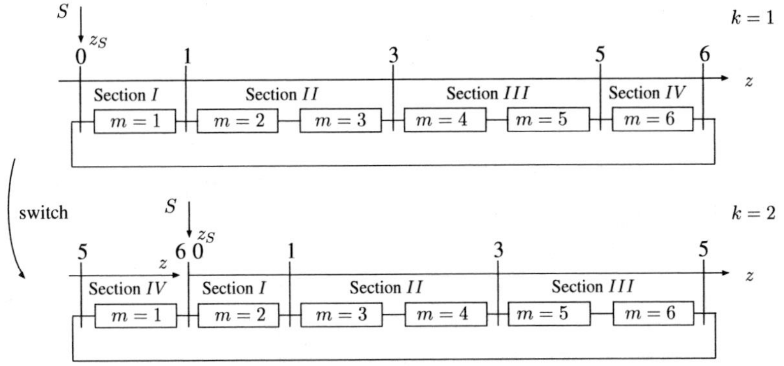

Figure 3.18: Sequence of port positions of a six column SMB plant (moving spatial coordinate)

Moving spatial coordinate z. Using a moving spatial coordinate, the origin of z is moved with each port switching and the value $z_S = 0$ remains the same (Figure 3.18):

$$z_S(k) = 0 \text{ for all } k\,.$$

Therefore, using the concept of the moving spatial coordinate yields one single discrete state for the discrete SMB subsystem for all k:

$$\mathsf{z}(k) = \mathsf{z}(0) = \begin{pmatrix} 0 \\ z_{rel,A} \\ z_{rel,A+B} \\ z_{rel,B} \end{pmatrix}.$$

Figure 3.19 shows the automaton graph of this system. With each switching event, the system returns to its discrete initial state. This automaton graph is independent of the number of columns n_c.

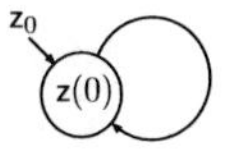

$$\mathsf{v} = \begin{pmatrix} 1 \\ 1 \\ 1 \\ 1 \end{pmatrix}$$

Figure 3.19: Example of an automaton graph of the discrete SMB subsystem using a moving coordinate system z

Then, the transition relation is

$$G:\ \mathsf{z}' = \mathsf{z} \tag{3.66}$$

with the initial state z_0, and the output relation is

$$H: \ \mathsf{w} = \mathsf{z}. \tag{3.67}$$

Conclusion. The synchronous switching of SMB processes significantly simplifies the general model (3.62), (3.64) of the discrete SCC subsystem. The number of discrete inputs is $n_{\mathsf{w}} = 2$. In case of a fixed spatial coordinate the conditionally executed actions $A1$, $A2$ and $A3$ of the state relation (3.62) are omitted to obtain G. The number of discrete states and outputs is $n_{\mathsf{z}} = n_{\mathsf{w}} = n_c$. If a moving spatial coordinate is applied the discrete state remains unchanged in case of a transition of the discrete subsystem. Hence, only one discrete state $\mathsf{z} = \mathsf{z}_0$ occurs and the output is equal to the state $\mathsf{w} = \mathsf{z}_0$.

3.5.3 Discrete–event subsystem of the VARICOL process

The modelling of the discrete subsystem of VARICOL processes is performed considering time–varying and time–invariant switching patterns. A time–varying switching pattern can occur if a port switching with respect to the propagation of the wave fronts is performed. In case of unknown disturbance inputs to the continuous subsystem, a suitable switching pattern can not be predicted and, hence, the control input to the discrete subsystem can not be determined in advance. For this case, the total number n_{z} of possible port position combinations on a circle of n_c separation columns is determined, for which Assumption 2.4.1 holds. Because a large number of discrete states and state transitions is obtained, the automaton is not presented for this case.

A time–invariant switching pattern reduces the complexity of the discrete subsystem significantly. The port switching is modelled for this case and analysed for both the fixed and the moving spatial coordinate system. It is shown how the complexity is further reduced applying the moving spatial coordinate.

Time–varying switching pattern. The application of the modelling concept presented in Section 3.5.1 to the representation of the discrete subsystem of VARICOL processes results in a more complex model than in case of the SMB process because all combinations $\mathsf{v}_i \in \{0, 1\}$, $i = 1, 2, 3, 4$ of the discrete input v are considered for all possible discrete states which fulfil the condition (3.63). In the general case, the switching pattern of the VARICOL process is time–varying, i.e. it is not guaranteed that a cyclic sequence of discrete inputs v is applied to the discrete subsystem of the VARICOL. However, the number of possible port position combinations on a circle of separation columns, which fulfil the condition (3.56), is limited. This number is determined by counting all possible port positions and their shifting on the circle of separation columns. The total of the possible port position combinations is a function of the number of columns n_c and corresponds to the total number $n_{\mathbf{z}}$ of discrete states. If a moving spatial coordinate is applied, the following expression for $n_{\mathbf{z}}$ is obtained:

$$n_{\mathbf{z}} = \left(\sum_{i=1}^{n_c} (n_c + 1)\, i \; - \; i^2 \right) . \tag{3.68}$$

For a theoretical VARICOL setup with three columns, this reveals

$$n_{\mathbf{z}}(n_c = 3) = 10$$

and for $n_c = 4$, $n_c = 5$ and $n_c = 6$

$$\begin{aligned} n_{\mathbf{z}}(n_c = 4) &= 20 \\ n_{\mathbf{z}}(n_c = 5) &= 35 \\ n_{\mathbf{z}}(n_c = 6) &= 56 . \end{aligned}$$

In case of a fixed spatial coordinate, the numbers of states have to be multiplied with n_c and the following results are obtained:

$$n_{\mathbf{z}} = \left(\sum_{i=1}^{n_c} (n_c + 1)\, i \; - \; i^2 \right) n_c , \tag{3.69}$$

$$\begin{aligned} n_{\mathbf{z}}(n_c = 3) &= 30 \\ n_{\mathbf{z}}(n_c = 4) &= 80 \\ n_{\mathbf{z}}(n_c = 5) &= 175 \\ n_{\mathbf{z}}(n_c = 6) &= 336 . \end{aligned}$$

The results show that in the general case the number of possible port position combinations is large. A representation by an automaton graph is therefore omitted. Obviously, if the time–varying switching pattern is considered, the discrete subsystem shows a considerable complexity compared to the discrete SMB subsystem.

For the description of the dynamics of the discrete subsystem of the VARICOL process considering a time–varying switching pattern, the transition relation G by Equation (3.62) and the output relation H by Equation (3.64) is applied without modifications, if a fixed spatial coordinate system is applied. In case of a moving spatial coordinate, the conditionally executed action $A4$ in Equation (3.62) has to be modified to obtain G:

$$
\begin{array}{llllll}
G: & & & & \mathsf{z}' = \mathsf{z} + \mathsf{v} & \\
& A1: & \text{if} \;\; \mathsf{z}_1' > \mathsf{z}_2' & \text{then set } \mathsf{z}_2' \text{ to} & \mathsf{z}_2' = \mathsf{z}_1' & \\
& & \text{if} \;\; \mathsf{z}_2' > \mathsf{z}_3' & \text{then set } \mathsf{z}_3' \text{ to} & \mathsf{z}_3' = \mathsf{z}_2' & \\
& A2: & \text{if} \;\; \mathsf{z}_3' \geq \mathsf{z}_4' & \text{then set } \mathsf{z}_4' \text{ to} & \mathsf{z}_4' = \mathsf{z}_3' + 1 & (3.70) \\
& A3: & \text{if} \;\; \mathsf{z}_4' - \mathsf{z}_1' \geq n_c & \text{then set } \mathsf{z}_1' \text{ to} & \mathsf{z}_1' = \mathsf{z}_1' + 1 & \\
& & & \text{and} & \text{repeat } A1 \text{ and } A2 & \\
& A4: & \text{if} \;\; \mathsf{z}_1' \geq 1 & \text{then set } \mathsf{z}' \text{ to} & \mathsf{z}' = \mathsf{z}' - 1\,. &
\end{array}
$$

Time–invariant switching pattern. If a time–invariant switching pattern is considered, there are four distinct discrete input variables, which are applied to the system in a cyclic manner. For the analysis, the following VARICOL process is considered:

- ◇ The initial column configuration is given by $(n_{c,0,I}/n_{c,0,II}/n_{c,0,III}/n_{c,0,IV})$ and fulfils the conditions (3.53) and (3.54).
- ◇ The switching times T_S, ΔT_A, ΔT_{A+B} and ΔT_B are given.

Example. The dynamical behaviour of this system is discussed in an example. If the solvent inlet port S is placed to the inlet of column $m = 1$, the initial state $\mathsf{z}(k_0)$ is determined based on the initial column configuration:

$$
\mathsf{z}(k_0) = \begin{pmatrix} 0 \\ n_{c,0,I} \\ n_{c,0,I} + n_{c,0,II} \\ n_{c,0,I} + n_{c,0,II} + n_{c,0,III} \end{pmatrix} .
$$

Hence, for an example with an initial column configuration of (0/1/2/1), the initial discrete state is

$$
\mathsf{z}(k_0) = \begin{pmatrix} 0 \\ 0 \\ 1 \\ 3 \end{pmatrix} .
$$

The sequence of input signals specifies the sequence of port switchings, which is determined by the values of the relative switching times. For the present analysis it is assumed that

$$\Delta T_{A+B} < \Delta T_A < \Delta T_B < T_S\,.$$

Then, $A + B$ is switched first, A is switched second, B is switched third and S is switched last during one period k_S. Hence, the sequence of discrete input signals during one period is

$$\begin{aligned}
\mathsf{v}(k_0) = \mathsf{v1} = \begin{pmatrix} 0 \\ 0 \\ 1 \\ 0 \end{pmatrix}, \quad \mathsf{v}(k_0+1) = \mathsf{v2} = \begin{pmatrix} 0 \\ 1 \\ 0 \\ 0 \end{pmatrix}, \\
\mathsf{v}(k_0+2) = \mathsf{v3} = \begin{pmatrix} 0 \\ 0 \\ 0 \\ 1 \end{pmatrix}, \quad \mathsf{v}(k_0+3) = \mathsf{v4} = \begin{pmatrix} 1 \\ 0 \\ 0 \\ 0 \end{pmatrix}.
\end{aligned} \tag{3.71}$$

Considering the initial state, the following sequence of states and outputs is obtained for the first switching period k_0:

$$\begin{aligned}
\mathsf{z}(k_0) = \mathsf{w}(k_0) = \begin{pmatrix} 0 \\ 0 \\ 1 \\ 3 \end{pmatrix}, \quad \mathsf{z}(k_0+1) = \mathsf{w}(k_0+1) = \begin{pmatrix} 0 \\ 0 \\ 2 \\ 3 \end{pmatrix}, \\
\mathsf{z}(k_0+2) = \mathsf{w}(k_0+2) = \begin{pmatrix} 0 \\ 1 \\ 2 \\ 3 \end{pmatrix}, \quad \mathsf{z}(k_0+3) = \mathsf{w}(k_0+3) = \begin{pmatrix} 0 \\ 1 \\ 2 \\ 4 \end{pmatrix}.
\end{aligned} \tag{3.72}$$

If a fixed spatial coordinate, which origin is placed to the inlet of column $m = 1$, is used, the subsequent states and outputs for $k_0 + 4$, $k_0 + 5$, $k_0 + 6$ and $k_0 + 7$ are obtained applying the input sequence (3.71):

$$\mathsf{z}(k_0+4)=\mathsf{w}(k_0+4)=\begin{pmatrix}1\\1\\2\\4\end{pmatrix},\qquad \mathsf{z}(k_0+5)=\mathsf{w}(k_0+5)=\begin{pmatrix}1\\1\\3\\4\end{pmatrix},$$

$$\mathsf{z}(k_0+6)=\mathsf{w}(k_0+6)=\begin{pmatrix}1\\2\\3\\4\end{pmatrix},\quad \mathsf{z}(k_0+7)=\begin{pmatrix}1\\2\\3\\5\end{pmatrix},\quad \mathsf{w}(k_0+7)=\begin{pmatrix}1\\2\\3\\1\end{pmatrix}.$$

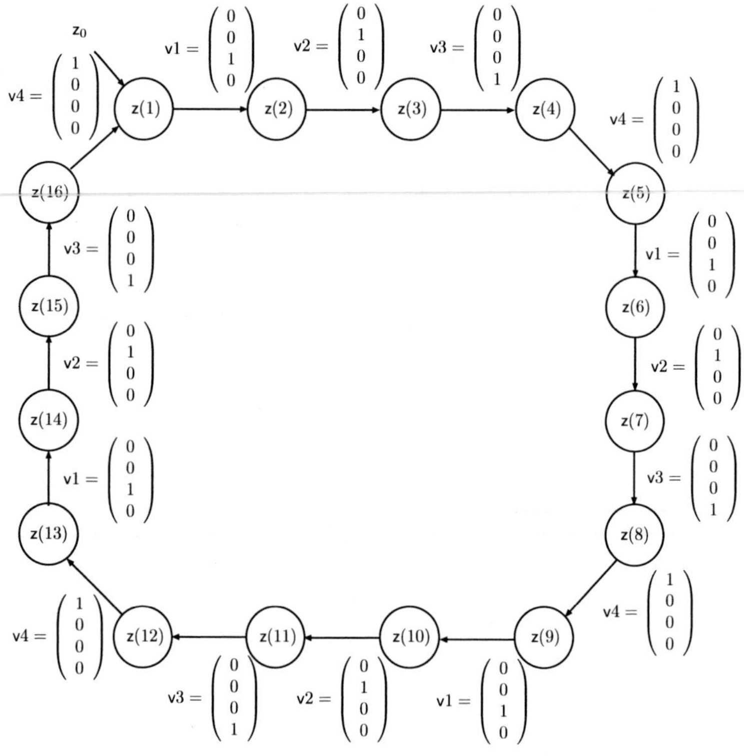

Figure 3.20: Example of an automaton graph of the discrete VARICOL subsystem for a time–invariant switching pattern using a fixed spatial coordinate system z

All subsequent states $\mathsf{z}(k)$ and outputs $\mathsf{w}(k)$ for $k=k_0+8, k_0+9, \ldots$ are determined by applying the input sequence (3.71) to the state transition relation G (Equation (3.62)) and the output relation

H (Equation (3.64)).

After $n_c = 4$ switching periods, i.e. after one switching cycle K, the state and the output reaches the initial state and output $\mathsf{z}(k_0)$ and $\mathsf{w}(k_0)$. In other words, if a fixed coordinate system is used, the VARICOL process, to which a time–invariant switching pattern is applied, has $n_\mathsf{z} = n_\mathsf{w} = 4 \cdot n_c$ discrete states or outputs, respectively. Figure 3.20 shows the automaton graph for this example with $k_0 = 1$.

If, by contrast, a moving spatial coordinate is used, which origin is tied to the position of the port S, the sequence of states and outputs corresponds to that of the first switching period (3.72), which was obtained for a fixed spatial coordinate system, with the exception that the fourth transition leads back to the initial state

$$\mathsf{z}(k_0+4) = \mathsf{w}(k_0+4) = \begin{pmatrix} 0 \\ 0 \\ 1 \\ 3 \end{pmatrix},$$

which is equal to the state and output at k_0. To consider this latter transition, Equation (3.70) has to be applied. The output relation (3.64) is applied without modification. Hence, using a moving coordinate system, the VARICOL process with a time–invariant switching pattern has $n_\mathsf{z} = n_\mathsf{w} = 4$ discrete states or outputs, respectively. Figure 3.21 shows the automaton graph for this example with $k_0 = 1$.

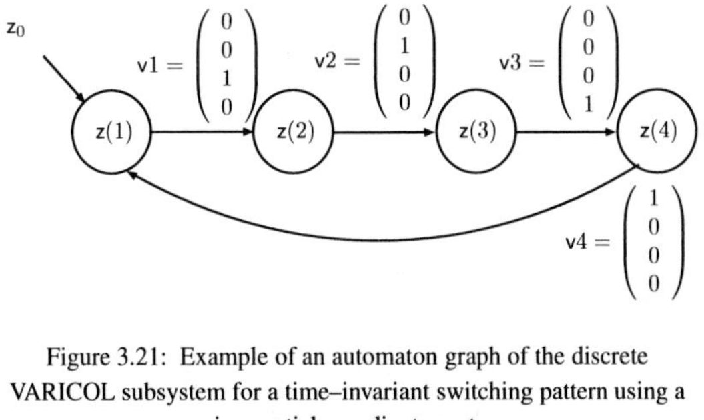

Figure 3.21: Example of an automaton graph of the discrete VARICOL subsystem for a time–invariant switching pattern using a moving spatial coordinate system z

Conclusion. The following results are obtained for the model of the discrete SCC subsystem: In case of a fixed spatial coordinate, the transition and output relation (3.62) and (3.64) are applied without modification. For a time–varying switching pattern, the number of states is given by

Equation (3.69), and for a time–invariant switching pattern, the number of discrete states is $n_z = 4\,n_c$ and a cyclic trajectory of z occurs. In case of a moving spatial coordinate the conditionally executed action $A4$ of the transition relation (3.62) has to be modified and Equation (3.70) is obtained. The output relation (3.64) remains unchanged. In case of a time–varying switching pattern, the number of states is given by Equation (3.68), and for the time–invariant switching pattern, the number of discrete states is $n_z = 4$ and z moves along a cyclic transition. Table 3.4 summarises the results of the analysis of the discrete SCC subsystems.

Switching pattern	Process	Spatial coordinate	n_z	G	H
time–varying	VARICOL	fixed	Eq. (3.69)	Eq. (3.62)	Eq. (3.64)
		moving	Eq. (3.68)	Eq. (3.70)	Eq. (3.64)
time–invariant	SMB	fixed	n_c	Eq. (3.65)	Eq. (3.64)
		moving	1	Eq. (3.66)	Eq. (3.67)
	VARICOL	fixed	$4\,n_c$	Eq. (3.62)	Eq. (3.64)
		moving	4	Eq. (3.70)	Eq. (3.64)

Table 3.4: Results of the analysis of the discrete SCC subsystems

3.6 Combined discrete and continuous dynamics

With the results of Sections 3.4 and 3.5, the combined discrete and continuous dynamics of SCC processes can be described.

As was shown in Section 3.4, the continuous dynamics depend upon the discrete output. Hence, a signal path from the discrete subsystem to the continuous subsystem via the injector $w(t) = I(\mathsf{w}(k))$ exists. However, the input of the discrete subsystem is not influenced by the continuous subsystem. Hence, the event generator E, which was shown in Figure 3.4, does not exist a priori in an SCC plant. For the operation of SCC plants, the event generator has to be designed as a discrete controller.

To analyse the coupling of the continuous and discrete subsystems, the hybrid automaton is used, which represents the discrete dynamics of a hybrid system by means of a finite automaton, and the continuous dynamics by means of differential equations. The following section presents the concept of the hybrid automaton. The representation of SMB processes by means of a hybrid automaton was first published in (Kleinert and Lunze, 2002). In this work, the concept is shown for the SMB in Section 3.6.2. It is extended to the VARICOL in Section 3.6.3 considering time–invariant switching patterns.

3.6.1 Hybrid automaton

To represent the interaction between the discrete and the continuous subsystem of a hybrid system, the concept of the hybrid automaton was introduced. A hybrid automaton can be interpreted as a generalisation of the timed automaton (Alur et al., 1993), in which the dynamical behaviour of continuous variables is represented in each automaton state by a set of differential equations. A hybrid automaton $\mathcal{A} = \{X, D, \mu_1, \mu_2, \mu_3\}$ can be represented by five elements:

1. the set $X \in \mathbb{R}^n$ of the continuous n-dimensional state $\boldsymbol{\zeta}$,
2. the set $\mathcal{D}$ of discrete states z (locations),
3. a labelling function μ_1 that assigns to each location z a system of differential equations which describe the dynamics of $\boldsymbol{\zeta}$,
4. a labelling function μ_2 assigning to each location z an exception set $\mu_2(\mathrm{z}) \subseteq X$. As long as $\boldsymbol{\zeta} \in X \backslash \mu_2(\mathrm{z})$ holds a transition of the location z is not possible, and
5. a labelling function μ_3 that assigns to each pair e of elements of D a transition relation $\mu_3(e) \subseteq X^2$. For $\mathrm{z}, \mathrm{z}' \in D$ and $\sigma, \sigma' \in X$ the state (σ', z') is called the successor state of (σ, z), and $(\sigma, \sigma') \in \mu_3(\mathrm{z}, \mathrm{z}')$.

The initial states are $\boldsymbol{\zeta}_0$ and z_0. The graphical representation of the hybrid automaton is called the automaton graph. Each location z is represented by a node, which is labelled with z, μ_1 and $\{X \backslash \mu_2\}$. Possible transitions are represented as edges of the graph which are labelled with the transition relation μ_3. Transition relations which assign the actual state to the successor state, like $\boldsymbol{\zeta} := \boldsymbol{\zeta}$ and $\mathrm{z} := \mathrm{z}$, are omitted.

3.6.2 Simulated Moving Bed process

SMB processes are synchronously switched SCC processes. The port switching is triggered using a clock

$$\dot{\tau} = 1 \, . \tag{3.73}$$

If the switching condition

$$\tau = T_S$$

is reached by the clock state τ, the port switching is performed and the clock state is reset to

$$\tau := 0\,.$$

Hence, in terms of the hybrid automaton, the continuous state of SMB processes consists of the distributed component concentrations $\boldsymbol{x}$ as described by Equation (3.48) and the clock state τ. $\boldsymbol{x}$ and τ are lumped to obtain the continuous state $\boldsymbol{\zeta}$ with

$$\boldsymbol{\zeta} = \begin{pmatrix} \boldsymbol{x} \\ \tau \end{pmatrix}. \tag{3.74}$$

Using the continuous clock state τ for the control of the port switching, a connection from the continuous subsystem to the discrete subsystem via the event generator

$$\begin{array}{llll} E(\tau): & \text{if} \quad \tau = T_S\,, & \text{then} & \mathsf{v} := \begin{pmatrix} 1 & 1 & 1 & 1 \end{pmatrix}' \\ & & \text{and} & \tau := 0 \\ & \text{else} & & \mathsf{v} := \begin{pmatrix} 0 & 0 & 0 & 0 \end{pmatrix}' \end{array} \tag{3.75}$$

is created.

Equation (3.75) shows that T_S can be considered as an input variable of the event generator. The block diagram of the time triggered system is shown in Figure 3.22.

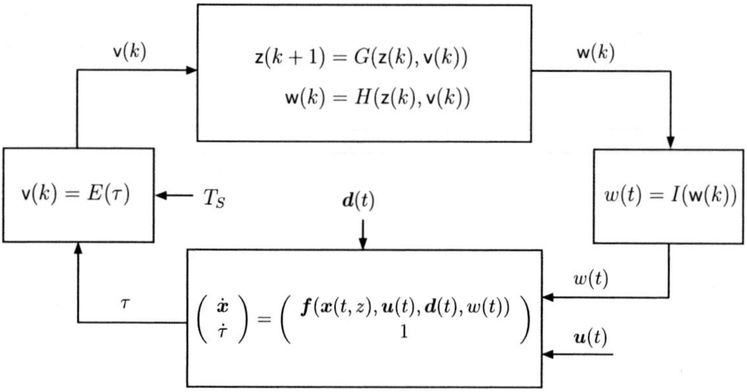

Figure 3.22: Block diagram of the time triggered SMB process

As was shown in Section 3.5.2, the complexity of the discrete SMB subsystem depends on whether a fixed or a moving spatial coordinate system z is chosen. To investigate the impact of the choice of z upon the overall system properties, this two approaches are treated separately in the following.

Fixed spatial coordinate system. The labelling function $\mu_1(\mathsf{z})$ which assigns to each discrete state z a set of possible continuous dynamics is

$$\mu_1(\mathsf{z}): \quad \dot{\boldsymbol{\zeta}} = \begin{pmatrix} \boldsymbol{f}(\boldsymbol{x}, \boldsymbol{u}, \boldsymbol{d}, w) \\ 1 \end{pmatrix}$$

according to the Equation (3.52), (3.73) and (3.74).

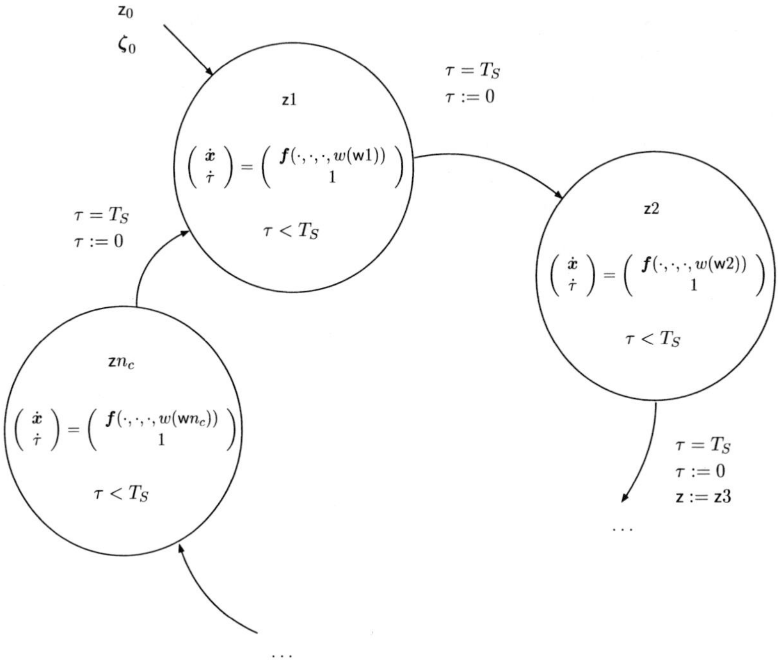

Figure 3.23: Hybrid automaton of SMB processes regarding a fixed spatial coordinate system z

In case of a fixed spatial coordinate system z the discrete SMB subsystem has $n_{\mathsf{z}} = n_{\mathsf{w}} = n_c$ discrete states and outputs. Considering the example discussed in Section 3.5.2 the following set D of locations is obtained:

$$\mathcal{D} = \{\mathsf{z}1, \mathsf{z}2, \ \ldots \ , \mathsf{z}n_c\}\,.$$

Correspondingly, a set of discrete outputs $\mathcal{O}$ can be given as

$$\mathcal{O} = \{\mathrm{w1}, \mathrm{w2}, \ \ldots \ , \mathrm{w}n_c\} .$$

The exception set μ_2 is determined by the event generator E introduced in Equation (3.75):

$$\mu_2(\mathrm{z}) = \{\tau | \ \tau = T_S\} .$$

If τ enters the exception set, a transition of z takes place.

The labelling function μ_3 is defined for the pairs $(\mathrm{z}, \mathrm{z}')$ of discrete state transitions which occur in the system:

$$(\mathrm{z1}, \mathrm{z2}) , \ (\mathrm{z2}, \mathrm{z3}) , \ \ldots \ , \ (\mathrm{z}n_c, \mathrm{z1}) .$$

$\mu_3(\mathrm{z}, \mathrm{z}')$ assigns to each discrete transition $(\mathrm{z}, \mathrm{z}')$ a successor state ζ' to the state ζ, which is here given by

$$\mu_3(\mathrm{z}, \mathrm{z}') : \ (\zeta, \zeta') = \left(\begin{pmatrix} \boldsymbol{x} \\ T_S \end{pmatrix} , \begin{pmatrix} \boldsymbol{x} \\ 0 \end{pmatrix} \right) .$$

Figure 3.23 shows partly the hybrid automaton graph of the system. The discrete transition starts at the initial state $\mathrm{z}_0 = \mathrm{z1}$ with the initial continuous state ζ_0 and follows along the states z2, z3 etc. until $\mathrm{z}n_c$ is reached. The successor state of $\mathrm{z}n_c$ is z1.

The hybrid automaton representation reveals three characteristic properties of the continuous subsystem of SMB processes:

1. The switching time T_S and the internal solvent mass flow rates $\boldsymbol{u}$ by Equation (3.50) are candidates for the control input.

2. Because for each discrete state different continuous dynamics $\boldsymbol{f}(\boldsymbol{x}, \boldsymbol{u}, \boldsymbol{d}, w)$ are assigned by μ_2, the SMB process with a fixed coordinate representation of the concentration profiles $\boldsymbol{x}$ is said to have *switching dynamics*.

3. Because the continuous state $\boldsymbol{x}$ does not change during a discrete transition, there are *no state jumps of the continuous state*.

Moving spatial coordinate system. Using a moving spatial coordinate z the same continuous state $\boldsymbol{\zeta}$

$$\boldsymbol{\zeta} = \begin{pmatrix} \boldsymbol{x} \\ \tau \end{pmatrix}$$

is used and the labelling function μ_1 is

$$\mu_1(\mathsf{z}) : \quad \begin{pmatrix} \dot{\boldsymbol{x}} \\ \dot{\tau} \end{pmatrix} = \begin{pmatrix} \boldsymbol{f}(\boldsymbol{x}, \boldsymbol{u}, \boldsymbol{d}, w) \\ 1 \end{pmatrix} .$$

If the variance of the separation column characteristics and the extra–column volumes is negligible according to Assumption 3.4.1, the change of the continuous dynamics during a discrete transition is negligible. Furthermore, the port positions do not change with respect to z. Therefore, the discrete subsystem has only one single location

$$\mathcal{D} = \{\mathsf{z1}\}$$

and one discrete output $\mathcal{O} = \{\mathsf{w1}\}$.

The exception set μ_2 is determined by Equation (3.75):

$$\mu_2(\mathsf{z}) = \{\tau | \; \tau = T_S\} .$$

When a transition of the discrete state occurs at $\tau = T_S$, the shifting of the origin of the spatial coordinate leads to an instantaneous change of the continuous state $\boldsymbol{x}$:

$$\begin{cases} \boldsymbol{x}'(0, z) = \boldsymbol{x}(T_S, z + L) , & z \in [0, (n_c - 1) \, L] \\ \boldsymbol{x}'(0, z) = \boldsymbol{x}(T_S, z - n_c \, L + 1) , & z \in [(n_c - 1) \, L, n_c \, L] , \end{cases} \tag{3.76}$$

where $\boldsymbol{x}'$ is the successor state of $\boldsymbol{x}$ in case of a discrete transition. Furthermore, the clock state τ is reset. Hence, the labelling function μ_3 is defined for the pair $(\mathsf{z}, \mathsf{z}') = (\mathsf{z1}, \mathsf{z1})$ only and assigns the transition of the continuous system by

$$\mu_3(\mathsf{z}, \mathsf{z}') : \quad (\boldsymbol{\zeta}, \boldsymbol{\zeta}') = \left(\begin{pmatrix} \boldsymbol{x} \\ T_S \end{pmatrix} , \begin{pmatrix} \boldsymbol{x}' \\ 0 \end{pmatrix} \right) .$$

$\boldsymbol{x}'$ is determined according to Equation (3.76).

The automaton graph consists of a single node with a self referring arc. It is shown in Figure 3.24.

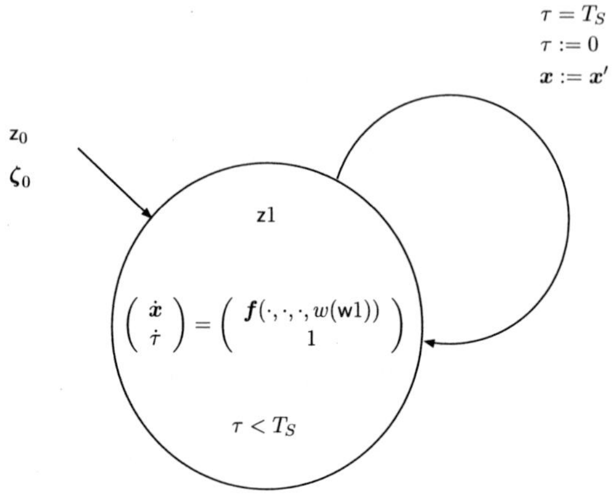

Figure 3.24: Hybrid automaton of SMB processes regarding a moving spatial coordinate system z

Three characteristic properties of the continuous subsystem of SMB processes which are described using a moving spatial coordinate z become obvious from the hybrid automaton representation:

1. The switching time T_S and the internal solvent mass flow rates $\boldsymbol{u}$ by Equation (3.50) are candidates for the control input.

2. Because the continuous dynamics $\boldsymbol{f}(\boldsymbol{x}, \boldsymbol{u}, \boldsymbol{d}, w)$ do not change according to Assumption 3.4.1 if a discrete transition occurs, the SMB process with a moving spatial coordinate system representation of the concentration profiles $\boldsymbol{x}$ has *no switching dynamics*.

3. A *jump of the continuous state* $\boldsymbol{x}$ occurs if a discrete transition takes place.

3.6.3 VARICOL process

The ports of the VARICOL process are switched asynchronously according to the switching pattern

$$(\Delta T_1, \Delta T_2, \Delta T_3, T_S)$$

for the initial column configuration

$$(n_{c,0,I}/n_{c,0,II}/n_{c,0,III}/n_{c,0,IV}),$$

where

$$\Delta T_1, \Delta T_2, \Delta T_3 \in \{\Delta T_A, \Delta T_{A+B}, \Delta T_B\}.$$

For the following analysis, it is assumed that the assignment of ΔT_1, ΔT_2 and ΔT_3 to the ports A, $A+B$ and B does not change and, hence, the switching pattern is time–invariant with respect to the sequence of the discrete states. Nevertheless, ΔT_1, ΔT_2, ΔT_3 and T_S may be time–varying. Using a clock with

$$\dot{\tau}_S = 1$$

the switching conditions are fulfilled if one or several of the following equations hold:

$$\begin{aligned} \tau_S &= \Delta T_1 \\ \tau_S &= \Delta T_2 \\ \tau_S &= \Delta T_3 \\ \tau_S &= T_S\,. \end{aligned}$$

In addition to the distributed state $\boldsymbol{x}$, τ_S is considered as a continuous state of the system and $\boldsymbol{\zeta}$ is

$$\boldsymbol{\zeta} = \begin{pmatrix} \boldsymbol{x} \\ \tau_S \end{pmatrix}.$$

If a switching condition is met, the corresponding switching signal v with

$$\mathsf{v} \in \left\{ \begin{pmatrix} 1 \\ 0 \\ 0 \\ 0 \end{pmatrix}, \begin{pmatrix} 0 \\ 1 \\ 0 \\ 0 \end{pmatrix}, \begin{pmatrix} 0 \\ 0 \\ 1 \\ 0 \end{pmatrix}, \begin{pmatrix} 0 \\ 0 \\ 0 \\ 1 \end{pmatrix} \right\}$$

is generated by the event generator E, which is

$$
\begin{aligned}
E(\tau_S): \quad & \text{if} \quad \tau_S = \Delta T_1\,, \quad \text{then} \quad \mathsf{v} := \mathsf{v1} \\
& \text{if} \quad \tau_S = \Delta T_2\,, \quad \text{then} \quad \mathsf{v} := \mathsf{v2} \\
& \text{if} \quad \tau_S = \Delta T_3\,, \quad \text{then} \quad \mathsf{v} := \mathsf{v3} \\
& \text{if} \quad \tau_S = T_S\,, \quad \text{then} \quad \mathsf{v} := \mathsf{v4}\,.
\end{aligned}
$$

The block diagram of this system is shown in Figure 3.25.

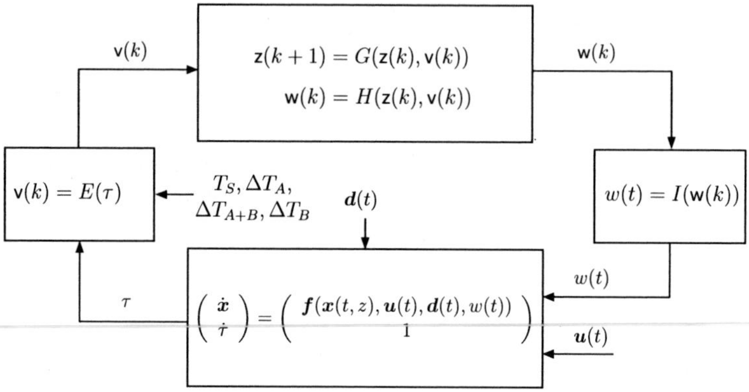

Figure 3.25: Block diagram of the time triggered VARICOL process

Like it was the case for the SMB process, the model complexity is different for the application of the fixed or the moving spatial coordinate. In the following, both cases are analysed separately.

Fixed spatial coordinate system. The labelling function μ_1 is

$$
\mu_1(\mathsf{z}): \quad \dot{\zeta} = \begin{pmatrix} \boldsymbol{f}(\boldsymbol{x}, \boldsymbol{u}, \boldsymbol{d}, w) \\ 1 \end{pmatrix}.
$$

Using a fixed spatial coordinate, the number of discrete states is $n_{\mathsf{z}} = 4\,n_c$. As an example, the four column VARICOL process presented in Section 3.5.3 is used. The set of locations for this example is

$$
\mathcal{D} = \{\mathsf{z1}\,,\ \mathsf{z2}\,, \ldots\,,\ \mathsf{z16}\}
$$

and the outputs are

$$
\mathcal{O} = \{\mathsf{w1}\,,\ \mathsf{w2}\,, \ldots\,,\ \mathsf{w16}\}\,.
$$

The exception set μ_2 is derived from the event generator E:

$$\begin{aligned}
\mu_2(\mathsf{z}) &= \{\tau_S |\ \tau_S = \Delta T_1\}\,, \quad \text{for } \mathsf{z} \in \{\mathsf{z1}\,,\ \mathsf{z5}\,,\ \mathsf{z9}\,,\ \mathsf{z13}\} \\
\mu_2(\mathsf{z}) &= \{\tau_S |\ \tau_S = \Delta T_2\}\,, \quad \text{for } \mathsf{z} \in \{\mathsf{z2}\,,\ \mathsf{z6}\,,\ \mathsf{z10}\,,\ \mathsf{z14}\} \\
\mu_2(\mathsf{z}) &= \{\tau_S |\ \tau_S = \Delta T_3\}\,, \quad \text{for } \mathsf{z} \in \{\mathsf{z3}\,,\ \mathsf{z7}\,,\ \mathsf{z11}\,,\ \mathsf{z15}\} \\
\mu_2(\mathsf{z}) &= \{\tau_S |\ \tau_S = T_S\}\,, \quad \text{for } \mathsf{z} \in \{\mathsf{z4}\,,\ \mathsf{z8}\,,\ \mathsf{z12}\,,\ \mathsf{z16}\}\,.
\end{aligned}$$

The labelling function μ_3, which assigns the successor state in case of a discrete transition, is

$$\mu_3(\mathsf{z}, \mathsf{z}') : \quad (\boldsymbol{\zeta}, \boldsymbol{\zeta}') = \left(\begin{pmatrix} \boldsymbol{x} \\ T_S \end{pmatrix}, \begin{pmatrix} \boldsymbol{x} \\ 0 \end{pmatrix} \right),$$

for the pairs

$$(\mathsf{z}, \mathsf{z}') \in \{(\mathsf{z4}, \mathsf{z5}), (\mathsf{z8}, \mathsf{z9}), (\mathsf{z12}, \mathsf{z13}), (\mathsf{z16}, \mathsf{z1})\}\,.$$

A part of the automaton graph of the VARICOL process, which is described using a fixed spatial coordinate system, is shown in Figure 3.26.

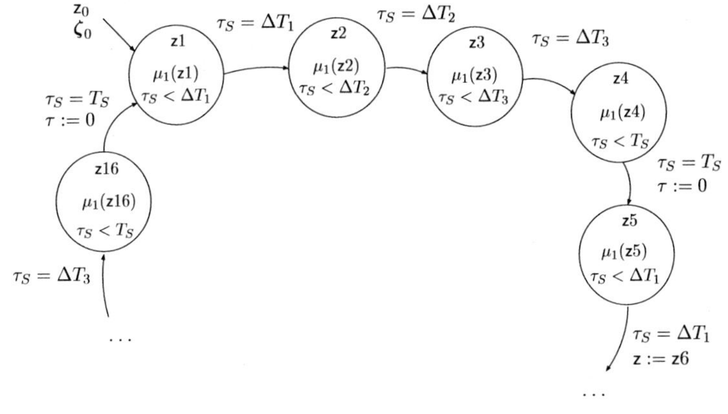

Figure 3.26: Example of a hybrid automaton of VARICOL processes using a fixed spatial coordinate system z

The trajectory of the discrete states starts with the initial state z1. Each single discrete state marks one subperiod κ. The sequence of the first four discrete states represents one switching period k_S. Such a sequence is performed n_c times before one switching cycle is complete. The successor state of the state z16 is the initial state z1. The switching cycle corresponds to the cycle K_S of the VARICOL process.

The hybrid automaton representation of the VARICOL process which is described using a fixed spatial coordinate system z shows the following properties of this process:

1. The free inputs to the process are the relative switching times ΔT_1, ΔT_2, ΔT_3, and the switching period duration T_S. These parameters determine the trajectory of the discrete subsystem. Further free inputs are the internal fluid flow rates, $\boldsymbol{u}$. These are possible control inputs to the process.

2. With each switching event, the labelling function μ_1 changes in the sense that the dynamics of the continuous subsystem, which describes the mass transfer in the plant, changes. Hence, the process shows *switching continuous dynamics*.

3. Because of the fixed spatial coordinate system there is *no jump of the continuous state* $\boldsymbol{x}$. Only the clock state τ_S is reset every fourth switching instant.

Moving spatial coordinate system. The continuous state variable ζ and the labelling function μ_1 remain unchanged:

$$\zeta = \begin{pmatrix} \boldsymbol{x} \\ \tau_S \end{pmatrix}, \quad \mu_1(\mathrm{z}) : \dot{\zeta} = \begin{pmatrix} \boldsymbol{f}(\boldsymbol{x}, \boldsymbol{u}, \boldsymbol{d}, w) \\ 1 \end{pmatrix}.$$

With each port switching, the port positions change with respect to the spatial coordinate z. Because of the cyclic nature of the switching pattern, the same port positions with respect to z occur after one switching period k_S. Therefore, if the difference of the separation column characteristics and the extra–column volumes are negligible according to Assumption 3.4.1, four discrete states z and four discrete outputs w have to be considered such that

$$\mathcal{D} = \{\mathrm{z1}, \mathrm{z2}, \mathrm{z3}, \mathrm{z4}\}$$

and

$$\mathcal{O} = \{\mathrm{w1}, \mathrm{w2}, \mathrm{w3}, \mathrm{w4}\}$$

is obtained. The exception sets for the continuous state τ_S are the same as for the fixed spatial coordinate for z1, z2, z3 and z4:

$$\mu_2(\mathsf{z}) = \{\tau_S |\ \tau_S = \Delta T_1\}\,, \quad \text{for } \mathsf{z} = \mathsf{z1}$$
$$\mu_2(\mathsf{z}) = \{\tau_S |\ \tau_S = \Delta T_2\}\,, \quad \text{for } \mathsf{z} = \mathsf{z2}$$
$$\mu_2(\mathsf{z}) = \{\tau_S |\ \tau_S = \Delta T_3\}\,, \quad \text{for } \mathsf{z} = \mathsf{z3}$$
$$\mu_2(\mathsf{z}) = \{\tau_S |\ \tau_S = T_S\}\,, \quad \text{for } \mathsf{z} = \mathsf{z4}\,.$$

When a transition due to the switching condition $\tau_S = T_S$ occurs, the spatial coordinate system is shifted by one column length. Hence, the continuous state $\boldsymbol{x}$ has to be restructured according to

$$\begin{cases} \boldsymbol{x}'(0,z) = \boldsymbol{x}(T_S, z + L)\,, & z \in [0, (n_c - 1)\, L] \\ \boldsymbol{x}'(0,z) = \boldsymbol{x}(T_S, z - n_c\, L + 1)\,, & z \in [(n_c - 1)\, L, n_c\, L]\,. \end{cases} \tag{3.77}$$

Furthermore, the clock state τ_S is reset. Hence, the labelling function μ_3 is

$$\mu_3(\mathsf{z}, \mathsf{z}') : \quad (\zeta, \zeta') = \left(\begin{pmatrix} \boldsymbol{x} \\ T_S \end{pmatrix}, \begin{pmatrix} \boldsymbol{x}' \\ 0 \end{pmatrix} \right),$$

for the pair $(\mathsf{z}, \mathsf{z}') = (\mathsf{z4}, \mathsf{z1})$. The successor state $\boldsymbol{x}'$ of $\boldsymbol{x}$ is determined according to Equation (3.77).

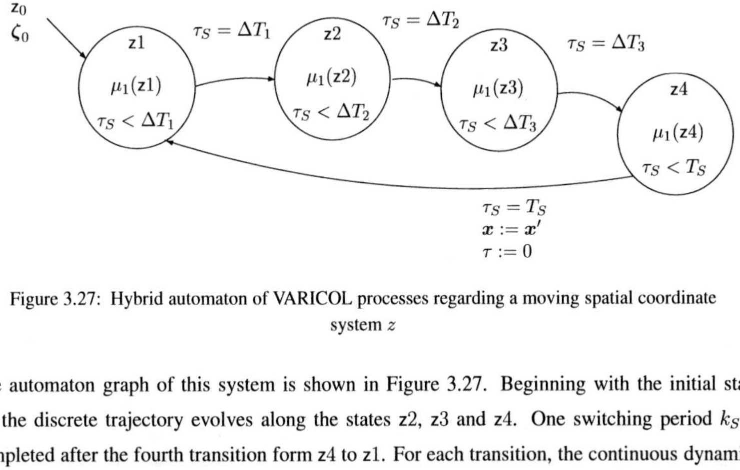

Figure 3.27: Hybrid automaton of VARICOL processes regarding a moving spatial coordinate system z

The automaton graph of this system is shown in Figure 3.27. Beginning with the initial state z1, the discrete trajectory evolves along the states z2, z3 and z4. One switching period k_S is completed after the fourth transition form z4 to z1. For each transition, the continuous dynamics change. The hybrid automaton representation reveals the following properties for the VARICOL process, to which the moving spatial coordinate system concept is applied:

1. The free inputs to the process are the relative switching times ΔT_1, ΔT_2, ΔT_3, and the switching period duration T_S. These parameters determine the trajectory of the discrete

VARICOL subsystem. Further free inputs are the internal fluid flow rates, $\boldsymbol{u}$. These are possible control inputs to the process.

2. With each switching event, the labelling function μ_1 changes in the sense that the dynamics of the continuous VARICOL subsystem changes. Hence, the process shows *switching continuous dynamics*.

3. Because the spatial coordinate system is shifted during the discrete transition $(\mathrm{z4}, \mathrm{z1})$, a *jump of the continuous state* $\boldsymbol{x}$ occurs. Additionally, the clock state τ_S is reset. Hence, applying the moving spatial coordinate system concept leads to both switching dynamics of the continuous subsystem *and* jumps of the continuous states.

Remark. If the assignment of the relative switching times ΔT_1, ΔT_2 and ΔT_3 to the ports A, $A + B$ and B changes in time, the automaton graphs still show cyclic transitions of the discrete state. However, each of these cycles might be different because differing transitions can occur in the switching periods of the cycles.

Remark. If the continuous dynamics of one separation column is represented by a lumped–parameter state–space model, the continuous SCC subsystem can be represented by one lumped–parameter state–space model (see e.g. (Ludemann-Homburger et al., 2000; Abel et al., 2002)). The hybrid automaton representation shows that deriving the solution of the state–space model for the moment of port switching, the SMB and the VARICOL with a time–invariant switching pattern can be represented by a discrete–time state–space model, of which the state represents the continuous state of the SCC in the moment of port switching.

3.7 Summary

This chapter presents the physical modelling of SCC processes. The distinct modelling of the continuous and the discrete parts of the overall hybrid dynamical system was performed.

The Equilibrium–Dispersive–Model for the a single separation column was derived. The procedure was applied to the TMB considering a close–loop separation column. An analytic solution of the stationary TMB model was derived. For the continuous subsystem of SCC processes, the complete model combining the singel–column models in terms of the node balances of the column interconnections was derived. A new representation using a block diagram was introduced. The complete continuous dynamics were represeted in a compact form for the use within the hybrid automaton representation.

The discrete SCC subsystem was modelled by a deterministic discrete–event system, regarding time–varying and time–invariant switching patterns as well as fixed and moving spatial coordinates. This is a new approach which allows for the explicit determination of the discrete state and output trajectory for a given port switching sequence of the SMB and the VARICOL.

The overall system dynamics of SCC processes considering time–invariant switching patterns and fixed and moving spatial coordinates was modelled and analysed by means of the hybrid automaton. As a result it was found that the event generator has to be designed as a discrete controller of the SCC plant. The free continuous inputs are the switching times and the internal fluid flow rates. These variables are candidates for the control input signal.

It was found that the discrete SCC subsystem shows different complexities depending upon the modelling concept with respect to the spatial coordinate and the switching pattern. Furthermore, the overall system dynamics show different hybrid system's phenomena, namely switching dynamics and jumps of the continuous states, if fixed or moving spatial coordinates, or the SMB or the VARICOL principle are applied.

Chapter 4

Wave front reconstruction

In case of steep wave fronts a reconstruction of the concentration profile wave fronts is necessary to obtain satisfactory control results. If only selected discrete–time concentration measuremets are available, a strongly simplified observer model should be used for this purpose. The herein presented approach is based on an explicit functional description of the wave fronts. The reconstruction is performed by the determination of the parameters of the functional description for each wave front and is referred to as the wave front observation. Based on the analysis of the observation error the quality of the wave front observation is investigated.

4.1 Observation problem

For the control of Simulated Counterflow Chromatographic (SCC) separation processes it is necessary to provide the controlled variables, which are the impurity concentrations of the recycling stream and the product purity values, to the controller. Principally, it is possible to directly measure all necessary controlled variables of the SCC process by discrete–time measurements (see Sections 1.2, 2.5). However, if steep wave fronts occur, e.g. due to nonlinear adsorption or low diffusion coefficients, the impact of disturbances can lead to a rapid change of the wave front concentrations in the recycling stream and in the product outlets. Hence, if the wave fronts c_1, c_2, c_3 and c_4 are available from a reconstruction, this disturbance impact can be determined in an early state and the control offset of the closed control loop, which occurs due to the disturbance, can be reduced. Because the purity values can directly be measured in each switching period, it is not necessary to determine the complete concentration profiles. The reconstruction of the wave fronts c_1, c_2, c_3 and c_4 from the input and output signals of the SCC process is sufficient. In the following, this reconstruction is referred to as the wave front observation.

It is desirable to provide a wave front observation applicable to both types of SCC processes, i.e. for synchronous and asynchronous port switching as well as for time–invariant and time–varying switching patterns. In this chapter, the general approach to the wave front modelling and observation of SCC processes based on discrete–time concentration measurements is presented. The results, which were previously published in (Kleinert and Lunze, 2004, 2005), are described in detail considering SMB processes with a constant switching time T_S. Because each wave front is considered as an independent system, the results can principally be applied to VARICOL processes with a time–invariant switching pattern.

The following **observation task** is considered:

Task 4.1.1 *Given is an SMB process with discrete–time concentration measurements* $c_A(\tau_m(k), z_m, k)$ *and* $c_B(\tau_m(k), z_m, k)$ *in each SMB section and a physical model based on the convection–diffusion equation which describes the dynamical behaviour of the concentration profiles. Determine the wave fronts* $c_1(\tau, z, k)$, $c_2(\tau, z, k)$, $c_3(\tau, z, k)$ *and* $c_4(\tau, z, k)$ *for all* $\tau \in [0, T_S]$, $z \in [0, L]$ *and* $k \in \mathbb{N}$. □

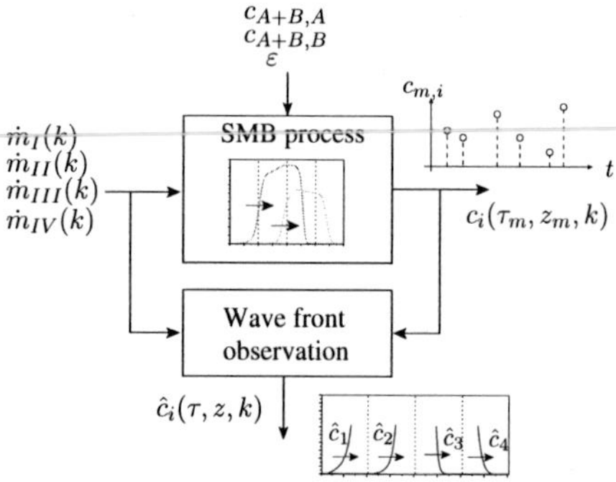

Figure 4.1: SMB wave front observation

Figure 4.1 shows the block diagram with the SMB process and the wave front observer.

The main problems with respect to the solution of Task 4.1.1 arise from the complexity of the physical model, the uncertainty of the model parameters, the complex process dynamics and the limited measurement information of the concentration profiles. Because it is supposed that only few measurements are available, the main information about the wave fronts have to be regarded implicitly in the model structure. Therefore, the observer model which is used in the herein presented approach is based on an explicit functional description of the form and the propagation of the wave fronts. The parameters of the functional description are determined by a state–observer.

The solution of Task 4.1.1 is structured as follows: In Section 4.2 the observer model is derived based on an explicit functional description of the SMB wave fronts. The parameters of the model are not known a priori and have to be determined from discrete–time concentration measurements. The observation task is formulated in terms of the wave front model and the solution is presented in Section 4.3. Section 4.4 proposes an approach to the analysis of the observation error which occurs due to model uncertainties and measurement errors.

4.2 Wave front modelling

The modelling purpose is to provide a model, which allows for the reconstruction of the SMB wave fronts from selected discrete–time concentration measurements. For the derivation of the observer model it has to be considered that, on the one hand, the model describes the SMB wave fronts with sufficient accuracy and, on the other hand, allows for the application of an observation principle which yields a simple observation algorithm as a solution of Task 4.1.1. For a better understanding of the modelling task and the model derivation the way of solution of the observation task is now briefly described.

Based on the approximation of the SMB by the TMB a model for the description of the *form* and the *movement* of the SMB wave fronts is derived. The parameters of the wave front models depend on the operation parameters of the SMB process and on the adsorption behaviour of the components A and B of the feed mixture and can not be determined in advance. A simple linear discrete–time state–space model is derived which describes the dynamics of the parameters. This state–space model allows for the application of a state observer to determine the model parameters independently of an initial estimate. Using these parameters with the model equation for the wave fronts c_i, $i = 1, 2, 3, 4$ allows to determine the values of the wave front concentrations $c_i(\tau, z, k)$ over the time $\tau \in [0, T_S]$ and the spatial range $z \in [0, L]$ in each SMB section.

For the derivation of the wave front model it is assumed that a constant switching time T_S is applied. The modelling principle is discussed in the following section. The model equations are derived in Section 4.2.2.

4.2.1 Modelling principle

As discussed in Section 4.1 it is sufficient to know the shape, position and movement of the wave fronts for the control. This idea yields the following proposition of the **wave front modelling task**:

Task 4.2.1 *Find for each SMB wave front c_i, $i = 1, 2, 3, 4$ a function μ with*

$$c_i'(\tau, z, k) = \mu(\tau, z, \boldsymbol{x}_i(k)) \,, \quad \tau \in [0, T_S] \,, \quad z \in [0, L] \,, \tag{4.1}$$

which describes the shape, position and movement of the wave front c_i for the switching period k such that

$$c_i'(\tau, z, k) \approx c_i(\tau, z, k)\,.$$

The parameter vector $\boldsymbol{x}_i(k)$ of the function μ shall be constant during one switching period but shall underlie a dynamical change in the transient state of the SMB process. Define the parameters $\boldsymbol{x}_i(k)$ as the states of the wave front and find a discrete–time state–space model $(\boldsymbol{f}, g)$ with

$$\begin{aligned} \boldsymbol{x}_i(k+1) &= \boldsymbol{f}(\boldsymbol{x}_i(k), \boldsymbol{u}_i(k), k) \\ y_i(k) &= g(\mu(\tau_{m,i}(k), z_{m,i}, \boldsymbol{x}_i(k)))\,, \end{aligned} \tag{4.2}$$

which describes the dynamics of $\boldsymbol{x}_i(k)$ in dependence upon the input $\boldsymbol{u}_i(k)$ of the SMB process, and which provides the output $y_i(k)$ as a measure of the wave front concentration measurement at the time $\tau_{m,i}(k)$ and the measurement position $z_{m,i}$. In the general case, the input $\boldsymbol{u}_i(k)$ compounds the control input of the SMB process which manipulates the wave front c_i. i.e. the internal fluid flow in the SMB section j corresponding to $z \in [0, L]$ and the switching time of the ports. □

The model proposed in Task 4.2.1 consists of the explicit functional description μ of a wave front and the state space model $(\boldsymbol{f}, g)$. Together with the state, input and output variables $\boldsymbol{x}_i$, $\boldsymbol{u}_i$ and y_i it forms a system which is considered as the wave front model:

Definition 4.2.1 *The system*

$$\mathcal{W}_i = \mathcal{W}_i\left(\ \boldsymbol{x}_i(k),\ \boldsymbol{u}_i(k),\ y_i(k),\ \mu,\ \boldsymbol{f},\ g(\mu)\ \right)$$

is called a WAVE FRONT MODEL. *For a wave front c_i of an SMB process the function μ describes the shape, position and the continuous propagation in dependence upon the local time τ, the spatial position z and the wave front state $\boldsymbol{x}_i(k)$ according to Equation (4.1). The dynamics of the wave front state $\boldsymbol{x}_i(k)$ is described by the function $\boldsymbol{f}$ and the output $y_i(k)$ is determined by the function g according to Equation (4.2).* □

The models for the four wave fronts differ only with respect to the values of the states $\boldsymbol{x}_i(k)$, whereas the model structure remains the same. Hence, the solution of Task 4.2.1 provides a model for each wave front c_i:

1. The wave front $c_1(\tau, z, k)$ is described by $c_1'(\tau, z, k)$. It propagates through the first column of the SMB section I (Figure 4.2). For the spatial representation the moving coordinate system z is used. $c_1'(\tau, z, k)$ describes the wave front in the range of $z \in [0, L]$, where $z = 0$ is the position z_S of the solvent inlet, and over the time horizon $\tau \in [0, T_S]$ of the switching period k. The position of the wave front with respect to z is influenced by the

fluid flow rate $\dot{m}_I(k)$ and the switching time T_S of the desorbent inlet position. If $\tau = T_S$, the solvent inlet port is switched by one column length, which implies the shifting of origin of z to the new port position. Because T_S is constant, $\dot{m}_I(k)$ is considered as the only input variable of the wave front c_1.

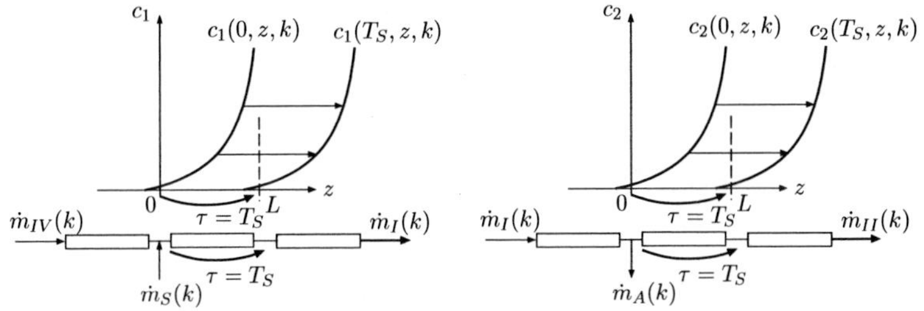

Figure 4.2: Wave front c_1 and c_2 (desorption wave fronts of the components A and B)

2. The wave front $c_2(\tau, z, k)$ is described by $c_2'(\tau, z, k)$. It propagates through the first column of the SMB section II (Figure 4.2). The moving spatial coordinate $z \in [0, L]$ is used. $c_2'(\tau, z, k)$ describes the wave front in z, where $z = 0$ is the position z_A of the outlet of A, and over the time horizon $\tau \in [0, T_S]$ of the switching period k. The position of the wave front with respect to z is influenced by the fluid flow rate $\dot{m}_{II}(k)$ and the switching of the outlet position of A. If $\tau = T_S$, the outlet port of A is switched by one column length, which implies the shifting of the origin of z to the new port position. Only $\dot{m}_{II}(k)$ is considered as an input variable of the wave front c_2.

3. The wave front $c_3(\tau, z, k)$ propagates through the last column of the SMB section III (Figure 4.3). The wave front is described by $c_3'(\tau, z, k)$ in the range $z \in [0, L]$ of the moving spatial coordinate, where $z = L$ is the position z_B of the outlet of component B, and over the time horizon $\tau \in [0, T_S]$ of the switching period k. The position of the wave front with respect to z is influenced by the fluid flow rate $\dot{m}_{III}(k)$ and the switching of the outlet position of the product B. If $\tau = T_S$, the outlet port of B is switched, which implies the shifting of the origin of z by one column length. $\dot{m}_{III}(k)$ is considered as input variable of the wave front c_3.

4. The wave front $c_4(\tau, z, k)$ propagates through the last column of the SMB section IV (Figure 4.3). It is described by $c_4'(\tau, z, k)$ using the moving spatial coordinate system. $c_4'(\tau, z, k)$ describes the wave front in the range of $z \in [0, L]$, where $z = L$ is the position of the recycling outlet which is the solvent inlet position z_S, and over the time horizon $\tau \in [0, T_S]$ of the switching period k. The position of the wave front with respect to z is influenced by the

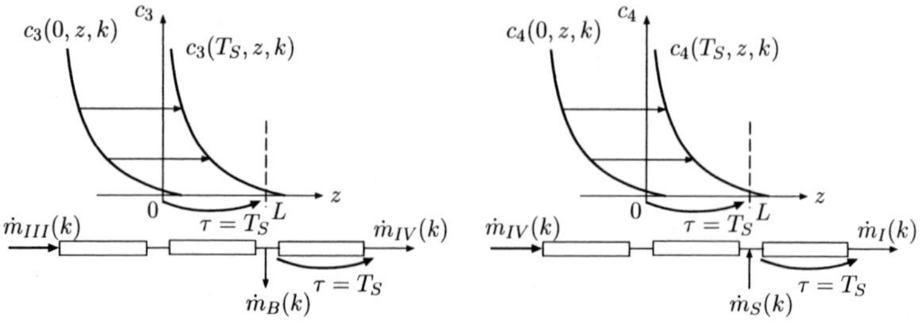

Figure 4.3: Wave front c_3 and c_4 (adsorption wave fronts of the components A and B)

fluid flow rate $\dot{m}_{IV}(k)$ and the switching of the solvent inlet. If $\tau = T_S$, the solvent inlet port is switched, which implies the shifting of the origin of z by one column length. The flow rate $\dot{m}_{IV}(k)$ is considered as the input variable of the wave front c_4.

It is supposed that for each SMB wave front one discrete–time and discrete–space concentration measurement is available per switching period k. This means that during one switching period one value of the wave front concentration $c_i(\tau, z_{m,i}, k)$ is recorded. The measurement time $\tau = \tau_{m,i}(k)$ can be different for each wave front c_i, and it can vary from one switching period to the next. The following measurement positions with respect to the position and propagation of the wave fronts are chosen:

1. Choose $z_{m,1} = z_S + 1$ for the wave front c_1.
2. Choose $z_{m,2} = z_A + 1$ for the wave front c_2.
3. Choose $z_{m,3} = z_B - 1$ for the wave front c_3.
4. Choose $z_{m,4} = z_S - 1$ for the wave front c_4.

Figure 4.4 shows where the measurement positions are located in an 8–column SMB plant with two columns per SMB section. Figure 4.5 shows the concentration measurement for the four wave fronts and gives an example for the selected measurements $c_i(\tau_{m,i}(k), z_{m,i}, k)$. The representation of the wave front measurement by the model, which is to be derived, is obtained from Equation (4.1) and is referred to as the *model output*:

$$c_i'(\tau_{m,i}(k), z_{m,i}, k) = \mu(\tau_{m,i}(k), z_{m,i}, \boldsymbol{x}_i(k)) .$$

The *output variable of the state–space model* of the wave front parameters is determined by the function g which is given by Equation (4.2):

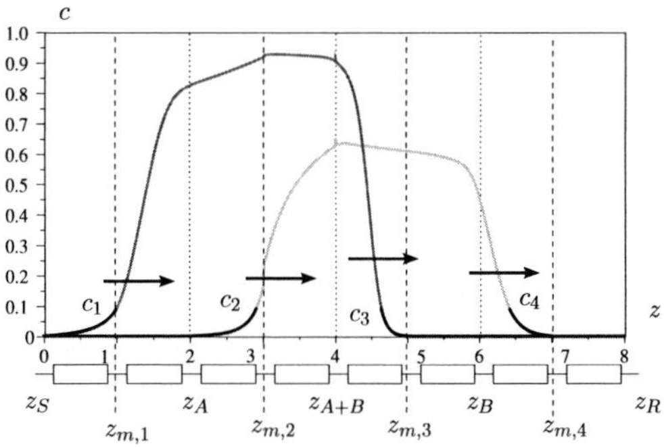

Figure 4.4: Measurement positions for the wave front reconstruction in an 8–column SMB plant

$$y_i(k) = g(\mu(\tau_{m,i}(k), z_{m,i}, \boldsymbol{x}_i(k))) \,.$$

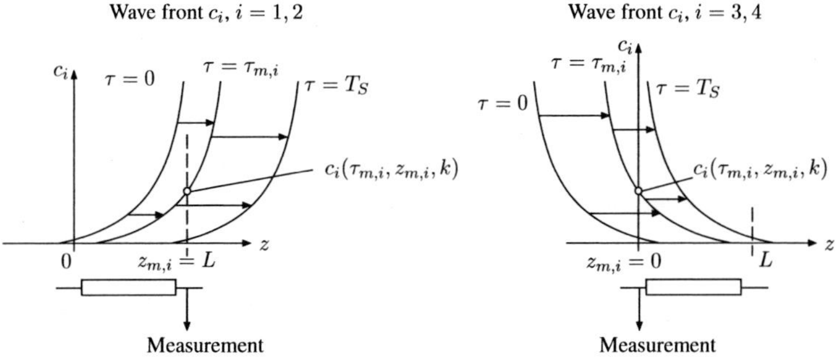

Figure 4.5: Selected measurements of the wave fronts

4.2.2 Wave front model

The derivation of the wave front model is performed considering an SMB operation mode, for which

- ◇ the wave front c_1 mainly propagates through the SMB section I,
- ◇ the wave front c_2 mainly propagates through the SMB section II,
- ◇ the wave front c_3 mainly propagates through the SMB section III,

◇ and the wave front c_4 mainly propagates through the SMB section IV.

The derivation is performed in four steps. In the first step the explicit functional description of the shape $c(\tau, z, k)$ of the wave front is derived. In the second step the parameters of the wave front model are defined. In the third step the dynamical behaviour of the parameters in the case of a transition of the SMB process are investigated. In the fourth step the functional description μ and the state vector $\boldsymbol{x}_i(k)$ of the wave front model are defined. Furthermore, the state–space model $(\boldsymbol{f}, g)$ for the description of the dynamics of $\boldsymbol{x}_i(k)$ and the output $y_i(k)$ is derived. Hereafter, the *autonomous linear wave front model* for SMB processes is introduced.

Step 1: Description of $c(\tau, z, k)$. To obtain a simple explicit functional description c' of the wave front shape in an SMB plant the stationary solution (3.36) of the closed–loop TMB model is applied, which yields

$$
\begin{aligned}
\text{Section } I: \quad & c_{stat,A,I}(z^\star) &&= r_{stat,A,I} \cdot \mathrm{e}^{\frac{v_{stat,A,I}}{D} z^\star} + s_{stat,A,I}\,, && z^\star \in [0,1]\\
\text{Section } II: \quad & c_{stat,B,II}(z^\star) &&= r_{stat,B,II} \cdot \mathrm{e}^{\frac{v_{stat,B,II}}{D} (z^\star-1)} + s_{stat,B,II}\,, && z^\star \in [1,2]\\
\text{Section } III: \quad & c_{stat,A,III}(z^\star) &&= r_{stat,A,III} \cdot \mathrm{e}^{\frac{v_{stat,A,III}}{D} (z^\star-2)} + s_{stat,A,III}\,, && z^\star \in [2,3]\\
\text{Section } IV: \quad & c_{stat,B,IV}(z^\star) &&= r_{stat,B,IV} \cdot \mathrm{e}^{\frac{v_{stat,B,IV}}{D} (z^\star-3)} + s_{stat,B,IV}\,, && z^\star \in [3,4]\,.
\end{aligned}
$$

The descriptions of the wave fronts in each SMB section use the same class of functions. Therefore, for the further derivation steps the following notation for $z \in [0, L]$ (where $L = 1$ is the normalised length of one separation column) is used:

$$
\begin{aligned}
\text{Wave front } 1: \quad & c'_1(z) &:=& \; c_{stat,A,I}(z^\star) & \Rightarrow \; & c'_1(z) := r_1\, \mathrm{e}^{\frac{v_1}{D} z} + s_1\\
\text{Wave front } 2: \quad & c'_2(z) &:=& \; c_{stat,B,II}(z^\star) & \Rightarrow \; & c'_2(z) := r_2\, \mathrm{e}^{\frac{v_2}{D} z} + s_2\\
\text{Wave front } 3: \quad & c'_3(z) &:=& \; c_{stat,A,III}(z^\star) & \Rightarrow \; & c'_3(z) := r_3\, \mathrm{e}^{\frac{v_3}{D} z} + s_3\\
\text{Wave front } 4: \quad & c'_4(z) &:=& \; c_{stat,B,IV}(z^\star) & \Rightarrow \; & c'_4(z) := r_4\, \mathrm{e}^{\frac{v_4}{D} z} + s_4\,.
\end{aligned}
$$

This yields

$$
c'_i(z) = r_i\, \mathrm{e}^{\frac{v_i}{D} z} + s_i \tag{4.3}
$$

for each wave front c_i, $i = 1, 2, 3, 4$.

The following transformations and simplifications are applied to Equation (4.3):

◇ Considering the results of the example process in Section 3.3.3 the parameter s_i is neglected.

- ◇ The parameter $\frac{v_i}{D}$ is determined by the relative flow velocity v_i (Equation (3.30)) and the diffusion coefficient D. The parameter is replaced using $b_i = \frac{v_i}{D}$.
- ◇ r_i is expressed as $r_i = \mathrm{e}^{-b_i\, z_{0,i}}$.

Applying these transformations and simplifications the following functional description of the wave front shape is obtained

$$\boxed{c_i'(z) = \mathrm{e}^{b_i\,(z - z_{0,i})}\,,} \tag{4.4}$$

where b_i is the shape parameter and $z_{0,i}$ is the spatial offset of the wave front. Equation (4.4) does not exactly represent the TMB wave fronts because the concentration offset described by s_i is neglected. With respect to the SMB it has to be considered that the concentration profile of the TMB process has approximately, but not exactly, the shape of the SMB process concentration profile. Therefore, Equation (4.4) is only an approximation of the shape of the SMB wave fronts. The following assumption is made:

Assumption 4.2.1 *The shape of the wave fronts in an SMB process can be described by Equation (4.4).* □

The impact of the diffusive effects which occur in real SMB plants is small. Therefore, the shape of the wave fronts is almost constant during one switching period k, and the spatial shift is assumed to be governed by a constant propagation velocity $v_{c,i}$. By applying the spatial shift $\Delta z(\tau) = -v_{c,i}\,\tau$ to Equation (4.4) the complete representation $c_i'(\tau, z, k)$ of the SMB wave front during the switching period k is obtained:

$$\boxed{c_i'(\tau, z, k) = \mathrm{e}^{b_i\,(z - (v_{c,i}\tau + z_{0,i}))}\,.} \tag{4.5}$$

Step 2: Parameter definition. The parameters b_i, $v_{c,i}$ and $z_{0,i}$ of Equation (4.5) determine the shape, the movement and the initial position at $\tau = 0$ of the wave front c_i. During the transition of the SMB process the shape, position and movement of the wave fronts change. This change can be tracked by a variation of the parameters b_i, $v_{c,i}$ and $z_{0,i}$. The following Assumption 4.2.2 is made:

Assumption 4.2.2 *The parameters b_i, $v_{c,i}$ and $z_{0,i}$ are constant during one switching period k but perform a dynamical change after each port switching if the SMB process performs a dynamical transition:*

$$b_i = b_i(k)\,, \quad v_{c,i} = v_{c,i}(k)\,, \quad z_{0,i} = z_{0,i}(k)\,.$$

□

A variation of the parameters $b_i(k)$, $v_{c,i}(k)$ and $z_{0,i}(k)$ is achieved by varying the flow rates $\dot{m}_j$ of the internal fluids in the SMB sections, through which the wave fronts propagate, or by a disturbance of the feed inlet concentration or the adsorption behaviour.

Step 3: Dynamical behaviour of the parameters. It is assumed that the propagation velocity $v_{c,i}$ of the wave front is constant in $z \in [0, L]$ and $\tau \in [0, T_S]$. Furthermore, the influence of the internal fluid flow onto the shape b_i and the propagation velocity is neglected for the modelling of the dynamical behaviour of the parameters. Therefore,

$$\begin{aligned} b_i(k+1) &= b_i(k) \\ v_{c,i}(k+1) &= v_{c,i}(k) \end{aligned} \tag{4.6}$$

is considered. In this case, the dynamics of the spatial offset only depends on the actual offset $z_{0,i}(k)$, the propagation velocity $v_{c,i}(k)$ and the switching time T_S:

$$z_{0,i}(k+1) = z_{0,i}(k) + T_S\, v_{c,i}(k) - L\,. \tag{4.7}$$

The substraction of L is necessary because of the shift of the spatial coordinate z in the moment of port switching. These considerations lead to the following assumption concerning the dynamics of the wave front parameters:

Assumption 4.2.3 *The parameters b_i and $v_{c,i}$ are assumed to be constant. Then, the dynamical behaviour of the spatial offset $z_{0,i}(k)$ only depends on the propagation velocity $v_{c,i}(k)$ and the actual switching time T_S according to Equation (4.7).* □

Step 4: Functional description μ and wave front state–space model $(\boldsymbol{f}, g)$. For the derivation of the states of the wave front model the factorisation

$$\boxed{\begin{aligned} x_{1,i}(k) &:= b_i(k) \\ x_{2,i}(k) &:= -b_i(k)\, v_{c,i}(k) \\ x_{3,i}(k) &:= -b_i(k)\, z_{0,i}(k) \end{aligned}} \tag{4.8}$$

is introduced, where $x_{1,i}(k)$, $x_{2,i}(k)$ and $x_{3,i}(k)$ are defined as the *states of the wave front model*:

$$\boldsymbol{x}_i(k) := \begin{pmatrix} x_{i,1}(k) \\ x_{i,2}(k) \\ x_{i,3}(k) \end{pmatrix} . \tag{4.9}$$

The functional description of the wave front $c_i(\tau, z, k)$ is obtained by applying Equation (4.8) and (4.9) to Equation (4.5):

$$\boxed{c_i'(\tau, z, k) = \mu(\tau, z, \boldsymbol{x}_i(k)) = \mathrm{e}^{(z \quad \tau \quad 1)\, \boldsymbol{x}_i(k)} .} \tag{4.10}$$

Applying Equations (4.8) and (4.9) to Equations (4.6) and (4.7) the state–space model

$$\boxed{\boldsymbol{x}_i(k+1) = \boldsymbol{f}(\boldsymbol{x}_i(k), k) ,} \tag{4.11}$$

with

$$\boxed{\boldsymbol{f}(\boldsymbol{x}_i(k)) = \underbrace{\begin{pmatrix} 1 & 0 & 0 \\ 0 & 1 & 0 \\ L & T_S & 1 \end{pmatrix}}_{\boldsymbol{A}} \boldsymbol{x}_i(k)} \tag{4.12}$$

of the dynamical behaviour of the states $\boldsymbol{x}_i(k)$ is obtained, which defines the function $\boldsymbol{f}$. Because $\boldsymbol{f}$ does not depend upon the control input, the resulting wave front state–space model is autonomous. Based on Equation (4.10) the output function g is derived. A suitable choice is the application of the natural logarithm to Equation (4.10). Hence, for

$$y_i(k) = g(\mu(\tau_{m,i}(k), z_{m,i}, \boldsymbol{x}_i(k))$$

the following function g

$$g(\mu(\tau_{m,i}(k), z_{m,i}, \boldsymbol{x}_i(k)) = \ln\left(\,\mu(\tau_{m,i}(k), z_{m,i}, \boldsymbol{x}_i(k))\,\right) \tag{4.13}$$

and thereby the output equation

$$y_i(k) = \underbrace{(z_{m,i} \;\; \tau_{m,i}(k) \;\; 1)}_{\boldsymbol{c}_i'(k)} \, \boldsymbol{x}_i(k) \tag{4.14}$$

is obtained.

Corollary 4.2.1 *The state $\boldsymbol{x}_i(k)$, the output $y_i(k)$ and the functions μ, $\boldsymbol{f}$ and g defined by the Equations (4.8) (4.9), (4.10), (4.12), (4.13) and (4.14) form a wave front model $\mathcal{W}_i = \mathcal{W}_i(\boldsymbol{x}_i, y_i, \mu, \boldsymbol{f}, g)$ for the wave fronts of an SMB process. $\boldsymbol{f}$ defines the autonomous linear dynamics of the state $\boldsymbol{x}_i(k)$. Because g is also a linear function of $\boldsymbol{x}_i(k)$ the wave front model $\mathcal{W}_i(\boldsymbol{x}_i, y_i, \mu, f, g)$ is referred to as an* AUTONOMOUS LINEAR WAVE FRONT MODEL. □

Solution 4.2.1 *The solution of Task 4.2.1 is given by the autonomous linear wave front model (Corollary 4.2.1)* □

4.3 Wave front observation

4.3.1 Formulation of the observation problem

In terms of the linear autonomous wave front model derived in the previous section, a new formulation of Task 4.1.1 is possible. It is assumed that the switching time T_S and the sequence of measurement times $\tau_{m,i}(k)$ of a wave front model are given. The dynamical behaviour of the considered wave front states is described by the autonomous linear time–varying discrete–time state–space model

$$\begin{aligned} \boldsymbol{x}_i(k+1) &= \boldsymbol{A}\,\boldsymbol{x}_i(k) \\ y_i(k) &= \boldsymbol{c}_i'(k)\,\boldsymbol{x}_i(k)\,, \end{aligned} \tag{4.15}$$

according to Equation (4.11), (4.12) and (4.14). The system has the unknown initial state $\boldsymbol{x}_i(0)$. The observer model is given by

$$\begin{aligned} \hat{\boldsymbol{x}}_i(k+1) &= \boldsymbol{A}\,\hat{\boldsymbol{x}}_i(k) \\ \hat{y}_i(k) &= \boldsymbol{c}_i'(k)\,\hat{\boldsymbol{x}}_i(k)\,, \end{aligned} \tag{4.16}$$

with an initial state $\hat{\boldsymbol{x}}_i(0) = \mathbf{0}$. The aim is to find a means to determine the state $\boldsymbol{x}_i(k)$ by the calculation of $\hat{\boldsymbol{x}}_i(k)$ under the consideration of the recent measurements $y_i(0), \ldots, y_i(k-1)$ and the recent model states $\hat{\boldsymbol{x}}_i(0), \ldots, \hat{\boldsymbol{x}}_i(k-1)$. The classical solution of this problem is to simulate the behaviour of $\hat{\boldsymbol{x}}_i(k)$ using Equation (4.16) and inject the observer output error $e_i(k) = y_i(k) - \hat{y}_i(k)$ via a feedback gain $\boldsymbol{l}_i$ to the state equation:

$$\begin{aligned} \hat{\boldsymbol{x}}_i(k+1) &= \boldsymbol{A}\,\hat{\boldsymbol{x}}_i(k) + \boldsymbol{l}_i\, e_i(k) \\ \hat{y}_i(k) &= \boldsymbol{c}_i'(k)\,\hat{\boldsymbol{x}}_i(k) \\ e_i(k) &= y_i(k) - \hat{y}_i(k)\,. \end{aligned} \tag{4.17}$$

The system represented by Equation (4.17) is called a Luenberger observer of the original system (4.15). The gain $\boldsymbol{l}_i$ is designed such that the observer state $\hat{\boldsymbol{x}}_i(k)$ is stable and converges to the state $\boldsymbol{x}_i(k)$ of the original system. These considerations lead to the formulation of the **observation problem** for the SMB wave fronts in terms of the following task:

Task 4.3.1 *Given is the output measurement $y_i(k)$ of an SMB wave front and the observer*

$$\begin{aligned} \hat{\boldsymbol{x}}_i(k+1) &= \boldsymbol{A}\,\hat{\boldsymbol{x}}_i(k) + \boldsymbol{l}_i\, e_i(k) \\ \hat{y}_i(k) &= \boldsymbol{c}_i'(k)\,\hat{\boldsymbol{x}}_i(k) \\ e_i(k) &= y_i(k) - \hat{y}_i(k) \end{aligned}$$

for the states of the wave front. Determine $\boldsymbol{l}_i$ such that the observer error $\tilde{\boldsymbol{x}}_i(k) = \boldsymbol{x}_i(k) - \hat{\boldsymbol{x}}_i(k)$ tends to zero, i.e.

$$\|\tilde{\boldsymbol{x}}_i(k)\|_{k\to\infty} \to 0\,.$$

□

Figure 4.6 represents the observer of an SMB wave front. Task 4.3.1 was formulated considering the observation of a single SMB wave front. For the observation of the four wave fronts of a binary SMB separation, the problem has to be solved for each of the four wave fronts. Because the considered wave front models have the same structure for each wave front and differ only with respect to the value of the parameters and the states, the solution of the observation problem will be presented for one wave front only. Therefore, the index i of the wave front will be omitted in the following section.

4.3.2 Observer design

For the design of the observer, the observability of the considered system is checked first. Then, the design of the observer is performed.

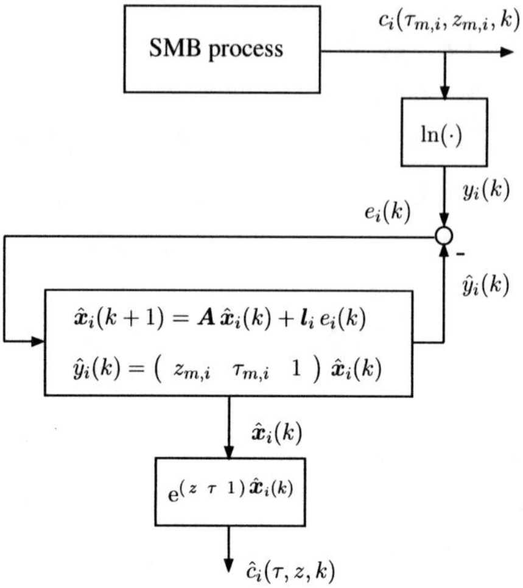

Figure 4.6: Block diagram of the wave front observer

Observability check for equidistantly sampled concentration measurements. Considering equidistantly sampled measurements with $\tau_m \neq \tau_m(k)$, a linear time-invariant state–space model for the dynamical behaviour of the wave front states is obtained

$$\hat{\boldsymbol{x}}(k+1) = \underbrace{\begin{pmatrix} 1 & 0 & 0 \\ 0 & 1 & 0 \\ L & T_S & 1 \end{pmatrix}}_{\boldsymbol{A}} \hat{\boldsymbol{x}}(k) \tag{4.18}$$

with a state dimension of $n_x = 3$. The output equation is

$$\hat{y}(k) = \underbrace{(z_m \;\; \tau_m \;\; 1)}_{\boldsymbol{c}'}\; \hat{\boldsymbol{x}}(k)\,. \tag{4.19}$$

A Luenberger observer, which the observer error $\tilde{\boldsymbol{x}}(k) = \boldsymbol{x}(k) - \hat{\boldsymbol{x}}(k)$ vanishes for $k \to \infty$, exists if the system given by Equation (4.18) and (4.19) is observable (Lunze, 2004b). The system is called observable if the rank of the observability matrix $\boldsymbol{S}_B$ is

$$\operatorname{rank}(\boldsymbol{S}_B) = n_x\,.$$

The observability matrix $\boldsymbol{S}_B$ of the considered system is

$$S_B = \begin{pmatrix} c' \\ c'\,A \\ c'\,A^2 \end{pmatrix} = \begin{pmatrix} z_m & \tau_m & 1 \\ z_m + L & \tau_m + T_S & 1 \\ z_m + 2\,L & \tau_m + 2T_S & 1 \end{pmatrix}.$$

It can be checked that

$$\mathrm{rank}(S_B) = 2 \neq n_x$$

holds, which means that the considered system is not observable. The interpretation of this result is that one state of the system is not observable. The result is not suprising because the considered state–space model (4.18) has two identical parallel systems which can be seen form the first two rows of the matrix A.

If, despite of the fact that the system is not observable, a classical Luenberger observer is applied to the observation of an SMB wave front in the stationary operation mode of the SMB, the observer output error $e(k) = y(k) - \hat{y}(k)$ vanishes, but the observer wave front has a different shape and propagation velocity compared to the SMB wave front. The fact that the observer output error vanishes means that the concentrations $c(\tau, z_m, k)$ and $\hat{c}(\tau, z_m, k)$ of the SMB wave front and the observer wave front coincide only at the measurement time $\tau = \tau_m$ (see. Figure 4.7).

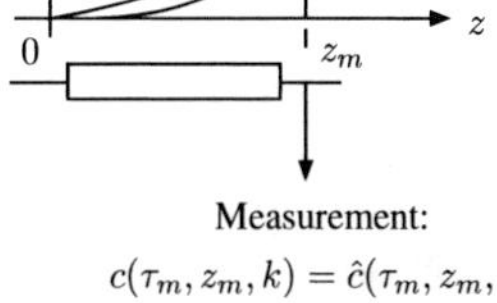

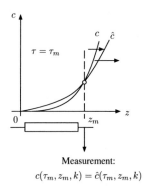

Figure 4.7: Observation result for a constant measurement time τ_m

Hence, applying equidistantly sampled measurements with $\tau_m \neq \tau_m(k)$, the solution of Task 4.3.1 is only possible if one wave front state is known exactly. Because none of the states are known a priori, a solution is only possible if a different way of concentration measurement is applied.

Observability check for non–equidistantly sampled concentration measurements. Because it is assumed that only one measurement is available per switching period, a measurement time

$$\tau_m = \tau_m(k)\,, \quad \tau_m(k) \neq \tau_m(k-1)\,,$$

which varies with each switching period, has to be considered to obtain observability. The following wave front state–space model is obtained:

$$\begin{aligned} \boldsymbol{x}(k+1) &= \underbrace{\begin{pmatrix} 1 & 0 & 0 \\ 0 & 1 & 0 \\ L & T_S & 1 \end{pmatrix}}_{\boldsymbol{A}} \boldsymbol{x}(k) \\ y(k) &= \underbrace{\begin{pmatrix} z_m & \tau_m(k) & 1 \end{pmatrix}}_{\boldsymbol{c}'(k)} \boldsymbol{x}(k)\,. \end{aligned} \tag{4.20}$$

Because of the time–varying output vector $\boldsymbol{c}'(k)$ the state–space model (4.20) is a linear time–varying (LTV) discrete–time system. A Luenberger observer, for which the observer error vanishes, exists if the system is observable. A criterion for the observability of autonomous linear discrete–time time–varying systems is given in (Ludyk, 1977). According to that criterion, a dynamical system $\mathcal{S}$ with

$$\mathcal{S} = \{\boldsymbol{x}(k), \boldsymbol{x}(k_0), \boldsymbol{y}(k), \boldsymbol{\Phi}(k, k_0), \boldsymbol{C}(k)\}$$

is considered, of which the dynamics of the state $\boldsymbol{x}(k) \in \mathbb{R}^{n_x}$ is described by the state equation

$$\boldsymbol{x}(k+1) = \boldsymbol{\Phi}(k, k_0)\, \boldsymbol{x}(k_0)\,,$$

where $\boldsymbol{\Phi}(k, k_0) = \prod_{i=0}^{k-k_0} \boldsymbol{A}(k_0 + i)$, $\boldsymbol{\Phi}(k_0, k_0) = \boldsymbol{A}(k_0)$ is the state transition matrix. The measured system output $\boldsymbol{y}(k) \in \mathbb{R}^{n_y}$ is given by the output equation

$$\boldsymbol{y}(k) = \boldsymbol{C}(k)\, \boldsymbol{x}(k)\,.$$

$\boldsymbol{A}(\cdot)$ or $\boldsymbol{C}(\cdot)$ are the time–varying system and output matrices of $\mathcal{S}$. The system $\mathcal{S}$ is called completely reconstructable after $k - k_0 = n_x$ steps, i.e. completely observable, if the rank of the observability matrix

$$\boldsymbol{S}_B(k,k_0) = \begin{pmatrix} \boldsymbol{C}(k_0) \\ \boldsymbol{C}(k_0+1)\,\boldsymbol{\Phi}(k_0,k_0) \\ \vdots \\ \boldsymbol{C}(k-1)\,\boldsymbol{\Phi}(k-2,k_0) \end{pmatrix}$$

is

$$\mathrm{rank}(\boldsymbol{S}_B(k,k_0)) = n_x \, .$$

Applied to the wave front state–space model given by Equation (4.20) the following observability matrix is obtained for $k_0 = 0$ and $k = 3$:

$$\boldsymbol{S}_B(3,0) = \begin{pmatrix} z_m & \tau_m(0) & 1 \\ z_m + L & \tau_m(1) + T_S & 1 \\ z_m + 2\,L & \tau_m(2) + 2\,T_S & 1 \end{pmatrix} .$$

If for all k the measurement time is varied such that

$$\tau_m(k) \neq \tau_m(k-1)$$

holds, then it can be shown that

$$\mathrm{rank}(\boldsymbol{S}_B(3,0)) = 3 \, .$$

Therefore, if e.g. an alternating measurement time $\tau_m(k)$ with

$$\boxed{\tau_m(k) \in \{\tau_{m1}, \tau_{m2}\} \, , \quad \tau_m(k) \neq \tau_m(k-1)} \tag{4.21}$$

is applied, the system (4.20) is observable.

Design of the observer. Because the wave front state–space model is an LTV system, the Luenberger observer design principle for LTV systems is used. The principle is based on the transformation of the original system into the observer canonical form and a design of an LTV Luenberger observer for the transformed system. The proportional feedback gain $\boldsymbol{l}$ of the observer is designed based on the choice of the dynamics of the observer canonical form of the original system. A detailed derivation of the design procedure is given in (Ludyk, 1981). Here, the derivation is shown for single–output systems.

Given is the discrete–time LTV system

$$\begin{aligned} \boldsymbol{x}(k+1) &= \boldsymbol{A}(k)\,\boldsymbol{x}(k) \\ y(k) &= \mathbf{c}'(k)\,\boldsymbol{x}(k) \end{aligned} \tag{4.22}$$

with the state variable $\boldsymbol{x} \in \mathbb{R}^{n_x}$ and the measured output $y \in \mathbb{R}$. $\boldsymbol{A}(k)$ and $\mathbf{c}'(k)$ are matrices or vectors of appropriate dimensions. The task is to design a proportional feedback gain $\boldsymbol{l}(k)$ such that the state $\hat{\boldsymbol{x}}(k)$ of the observer

$$\begin{aligned} \hat{\boldsymbol{x}}(k+1) &= (\,\boldsymbol{A}(k) - \boldsymbol{l}(k)\,\mathbf{c}'(k)\,)\,\hat{\boldsymbol{x}}(k) + \boldsymbol{l}(k)\,y(k) \\ \hat{y}(k) &= \mathbf{c}'(k)\,\hat{\boldsymbol{x}}(k) \end{aligned} \tag{4.23}$$

converges to the state $\boldsymbol{x}(k)$ of the system (4.22) and

$$\|\boldsymbol{x}(k) - \hat{\boldsymbol{x}}(k)\|_{k\to\infty} = 0\,.$$

The dynamical behaviour of the observer error $\tilde{\boldsymbol{x}}(k) = \boldsymbol{x}(k) - \hat{\boldsymbol{x}}(k)$ of the system (4.22), (4.23) is given by

$$\tilde{\boldsymbol{x}}(k+1) = \boldsymbol{H}(k)\,\tilde{\boldsymbol{x}}(k)\,, \tag{4.24}$$

with

$$\boldsymbol{H}(k) = \boldsymbol{A}(k) - \boldsymbol{l}(k)\,\mathbf{c}'(k)\,. \tag{4.25}$$

For a transformation of the system (4.22) to the observer canonical form, the state transformation

$$\boldsymbol{x}(k) = \boldsymbol{Q}(k)\,\boldsymbol{x}_B(k)\,, \tag{4.26}$$

where $\boldsymbol{x}_B(k)$ is the state of the transformed system, has to be applied. The resulting state–space model for $\boldsymbol{x}_B(k)$ is

$$\begin{aligned} \boldsymbol{x}_B(k+1) &= \underbrace{\boldsymbol{Q}^{-1}(k+1)\,\boldsymbol{A}(k)\,\boldsymbol{Q}(k)}_{\boldsymbol{A}_B(k)}\,\boldsymbol{x}_B(k) \\ y(k) &= \underbrace{\boldsymbol{c}'(k)\,\boldsymbol{Q}(k)}_{\boldsymbol{c}'_B(k)}\,\boldsymbol{x}_B(k)\,. \end{aligned} \tag{4.27}$$

The system matrix $\boldsymbol{A}_B(k)$ has the observer canonical form

$$\boldsymbol{A}_B(k) = \begin{pmatrix} 0 & 0 & \cdots & 0 & -a_0(k) \\ 1 & 0 & \cdots & 0 & -a_1(k) \\ 0 & 1 & \cdots & 0 & -a_2(k) \\ \vdots & & \ddots & & \vdots \\ 0 & \cdots & 0 & 1 & -a_{n_x-1}(k) \end{pmatrix},$$

where $a_i(k)$, $i = 1, 2, \ldots, n_x - 1$ are the coefficients of the characteristic polynomial and n_x is the state dimension.

The output vector $\boldsymbol{c}'_B(k)$ is time–invariant and has the special form

$$\boldsymbol{c}'_B(k) = \boldsymbol{c}'_B = (\,0 \quad \cdots \quad 0 \;\; 1\,)\,.$$

The transformation (4.26) exists for single–output LTV systems, if the system is completely observable. Then, the transformation matrix $\boldsymbol{Q}(k)$ is determined by

$$\begin{aligned} \boldsymbol{Q}(k) = \; & (\; \boldsymbol{s}_b(k) \quad \boldsymbol{A}(k-1)\,\boldsymbol{s}_b(k-1) \quad \boldsymbol{A}(k-1)\,\boldsymbol{A}(k-2)\,\boldsymbol{s}_b(k-2) \quad \cdots \\ & \boldsymbol{A}(k-1)\ldots\boldsymbol{A}(k-n_x+1)\,\boldsymbol{s}_b(k-n_x+1))\,. \end{aligned} \tag{4.28}$$

The column vector $\boldsymbol{s}_b(k)$ is the last column of the inverse of the observability matrix $\boldsymbol{S}_b(k)$, i.e.

$$\boldsymbol{s}_b(k) = \boldsymbol{S}_b^{-1}(k) \begin{pmatrix} 0 \\ \vdots \\ 0 \\ 1 \end{pmatrix},$$

with

$$S_b(k) = \begin{pmatrix} c'(k) \\ c'(k+1)\,A(k) \\ c'(k+2)\,A(k+1)\,A(k) \\ \vdots \\ c'(k+n_x-1)\,A(k+n_x-2)\cdots A(k) \end{pmatrix}.$$

Applying the transformation (4.26) to the observation error $\tilde{x}(k)$ of the original system, the dynamical behaviour of the observation error $\tilde{x}_B(k) = x_B(k) - \hat{x}_B(k)$ of the transformed system is obtained from Equation (4.24):

$$\tilde{x}_B(k+1) = \underbrace{Q^{-1}(k+1)\,H(k)\,Q(k)}_{H_{B(k)}}\,\tilde{x}_B(k). \tag{4.29}$$

The matrix $H_B(k)$ has the observer canonical form

$$H_B(k) = \begin{pmatrix} 0 & 0 & \cdots & 0 & -h_0(k) \\ 1 & 0 & \cdots & 0 & -h_1(k) \\ 0 & 1 & \cdots & 0 & -h_2(k) \\ \vdots & & \ddots & & \vdots \\ 0 & \cdots & 0 & 1 & -h_{n_x-1}(k) \end{pmatrix},$$

where $h_i(k)$ are the coefficients of the characteristic polynomial of the system (4.29).

With the Equations (4.25) and (4.29), the expression

$$Q(k+1)\,H_B(k) = A(k)\,Q(k) + l(k)\,\underbrace{c'(k)\,Q(k)}_{=\,c'_B}$$

is derived, which is transformed to

$$l(k)\,c'_B = A(k)\,Q(k) - Q(k+1)\,H_B(k).$$

Recalling the special form of the output vector $c'_B = (0 \ \cdots \ 0 \ 1)$, the observer feedback gain $l(k)$ is the last column of the matrix $A(k)\,Q(k) - Q(k+1)\,H_B(k)$. Hence, for $l(k)$ the following expression is obtained:

$$\boldsymbol{l}(k) = (\boldsymbol{A}(k)\,\boldsymbol{Q}(k) - \boldsymbol{Q}(k+1)\,\boldsymbol{H}_B(k)) \begin{pmatrix} 0 \\ \vdots \\ 0 \\ 1 \end{pmatrix} . \tag{4.30}$$

With Equation (4.30) the observer feedback gain $\boldsymbol{l}(k)$ can be designed by the choice of the error dynamics $\boldsymbol{H}_B(k)$: If $\boldsymbol{H}_B(k)$ is chosen such that

$$\lim_{k\to\infty} \|\tilde{\boldsymbol{x}}_B(k)\| = 0 \,,$$

then

$$\lim_{k\to\infty} \|\tilde{\boldsymbol{x}}(k)\| = 0$$

of the system (4.24) also holds and the observer given by Equations (4.23) and (4.30) is a solution of Task 4.3.1.

Application to the wave front observation. The observer for LTV systems is used for the observation of the SMB wave front states. For the observer design it is necessary to determine the transformation matrix $\boldsymbol{Q}(k)$ and $\boldsymbol{Q}(k+1)$ and to specify the observation error dynamics $\boldsymbol{H}_B(k)$ for the wave front state–space model

$$\begin{aligned} \boldsymbol{x}(k+1) &= \underbrace{\begin{pmatrix} 1 & 0 & 0 \\ 0 & 1 & 0 \\ L & T_S & 1 \end{pmatrix}}_{\boldsymbol{A}} \boldsymbol{x}(k) \\ y(k) &= \underbrace{\begin{pmatrix} z_m & \tau_m(k) & 1 \end{pmatrix}}_{\boldsymbol{c}'(k)} \boldsymbol{x}(k) \,. \end{aligned} \tag{4.31}$$

The observability matrix $\boldsymbol{S}_b(k)$ for this system is

$$\boldsymbol{S}_b(k) = \begin{pmatrix} \boldsymbol{c}'(k) \\ \boldsymbol{c}'(k+1)\,\boldsymbol{A} \\ \boldsymbol{c}'(k+2)\,\boldsymbol{A}^2 \end{pmatrix} = \begin{pmatrix} z_m & \tau_m(k) & 1 \\ z_m + L & \tau_m(k+1) + T_S & 1 \\ z_m + 2\,L & \tau_m(k+2) + 2\,L & 1 \end{pmatrix} .$$

The inverse of the observability matrix is

$$\boldsymbol{S}_b^{-1}(k) = \frac{1}{L\left(\tau_m(k) - 2\,\tau_m(k+1) + \tau_m(k+2)\right)} \cdot \begin{pmatrix} s_{b,11} & s_{b,12} & s_{b,13} \\ s_{b,21} & s_{b,22} & s_{b,23} \\ s_{b,31} & s_{b,32} & s_{b,33} \end{pmatrix},$$

with

$$\begin{aligned}
s_{b,11} &= \tau_m(k+1) - \tau_m(k+2) - T_S \\
s_{b,12} &= -\tau_m(k) + \tau_m(k+2) + 2\,T_S \\
s_{b,13} &= \tau_m(k) - \tau_m(k+1) - T_S \\
s_{b,21} &= 1 \\
s_{b,22} &= -2 \\
s_{b,23} &= 1 \\
s_{b,31} &= z_m\,\tau_m(k+2) + z_m\,T_S + L\tau_m(k+2) - z_m\,\tau_m(k+1) - 2\,L\tau_m(k+1) \\
s_{b,32} &= -z_m\,\tau_m(k+2) - 2\,z_m\,T_S + z_m\tau_m(k) + 2\,L\tau_m(k) \\
s_{b,33} &= z_m\,\tau_m(k+1) + z_m\,T_S - z_m\,\tau_m(k) - L\,\tau_m(k)\,.
\end{aligned}$$

The vector $\boldsymbol{s}_b(k)$ is the last column of $\boldsymbol{S}_b^{-1}(k)$:

$$\begin{aligned}
\boldsymbol{s}_b(k) &= \frac{1}{L\,(\tau_m(k)-2\,\tau_m(k+1)+\tau_m(k+2))} \cdot \ldots \\
&\ldots \begin{pmatrix} \tau_m(k) - \tau_m(k+1) - T_S \\ 1 \\ z_m\,\tau_m(k+1) + z_m\,T_S - z_m\,\tau_m(k) - L\,\tau_m(k) \end{pmatrix}.
\end{aligned}$$

With Equation (4.28) the transformation matrix $\boldsymbol{Q}(k)$ becomes

$$\boldsymbol{Q}(k) = \begin{pmatrix} q_{11} & q_{12} & q_{13} \\ q_{21} & q_{22} & q_{23} \\ q_{31} & q_{32} & q_{33} \end{pmatrix},$$

with

$$
\begin{aligned}
q_{11} &= \frac{-\tau_m(k)+\tau_m(k+1)+T_S}{L(-\tau_m(k)-\tau_m(k+2)+2\,\tau_m(k+1))}\\
q_{12} &= \frac{\tau_m(k-1)-\tau_m(k)-T_S}{L(\tau_m(k-1)+\tau_m(k+1)-2\,\tau_m(k))}\\
q_{13} &= \frac{-\tau_m(k-2)+\tau_m(k-1)+T_S}{L(-\tau_m(k-2)-\tau_m(k)+2\,\tau_m(k-1))}\\
q_{21} &= -\frac{1}{(-\tau_m(k)-\tau_m(k+2)+2\,\tau_m(k+1))}\\
q_{22} &= \frac{1}{(\tau_m(k-1)+\tau_m(k+1)-2\,\tau_m(k))}\\
q_{23} &= -\frac{1}{(-\tau_m(k-2)-\tau_m(k)+2\,\tau_m(k-1))}\\
q_{31} &= \frac{-z_m\,\tau_m(k+1)-z_m\,T_S+z_m\,\tau_m(k)+L\tau_m(k)}{L(-\tau_m(k)-\tau_m(k+2)+2\,\tau_m(k+1))}\\
q_{32} &= -\frac{L\tau_m(k)-z_m\,\tau_m(k)-z_m\,T_S+z_m\,\tau_m(k-1)}{L(\tau_m(k-1)+\tau_m(k+1)-2\,\tau_m(k))}\\
q_{33} &= \frac{-L\tau_m(k-2)+2\,L\tau_m(k-1)-z_m\,\tau_m(k-1)-z_m\,T_S+z_m\,\tau_m(k-2)}{L(-\tau_m(k-2)-\tau_m(k)+2\,\tau_m(k-1))}\,.
\end{aligned}
$$

$\boldsymbol{Q}(k+1)$ can directly be determined from $\boldsymbol{Q}(k)$. The elements of $\boldsymbol{Q}(k)$ where computed using the computer algebra–system MAPLE.

The next step is to chose $\boldsymbol{H}_B(k)$. For simplicity, $\boldsymbol{H}_B$ is chosen to be constant. As the considered wave front state–space model is a three dimensional system, $\boldsymbol{H}_B$ is a 3×3 matrix, and three coefficients h_0, h_1 and h_2 of the characteristic polynomial have to be chosen. A series of three first–order systems

$$
x_i(k+1) = a_i\,x_i(k) + b_i\,u_i(k)\,, \quad i = 1,2,3
$$

is considered as a reference system, of which the state–space model representation is

$$
\begin{pmatrix} x_1(k+1) \\ x_2(k+1) \\ x_3(k+1) \end{pmatrix} = \underbrace{\begin{pmatrix} a_1 & 0 & 0 \\ b_2 & a_2 & 0 \\ 0 & b_3 & a_3 \end{pmatrix}}_{\boldsymbol{A}_S} \begin{pmatrix} x_1(k) \\ x_2(k) \\ x_3(k) \end{pmatrix} + \begin{pmatrix} b_1 \\ 0 \\ 0 \end{pmatrix} u_1(k)\,.
$$

The characteristic polynomial of this system is

$$
\det(\lambda\boldsymbol{I} - \boldsymbol{A}_S) = \lambda^3 \underbrace{-(a_1+a_2+a_3)}_{h_2}\,\lambda^2 + \underbrace{a_1\,a_2 + a_1\,a_3 + a_2\,a_3}_{h_1}\,\lambda \underbrace{-a_1\,a_2\,a_3}_{h_0}\,.
$$

Now, the model and the design parameters have to be determined. An alternating measurement time

$$\tau_m(k) \in \{\tau_{m1}, \tau_{m2}\}\,, \quad \tau_m(k) \neq \tau_m(k-1)$$

is chosen. With this choice, only two observer gain vectors have to be designed. The following values are given for the wave front model:

$$\begin{aligned} T_S &= 120 \\ L &= 1 \\ z_m &= L \\ \tau_{m1} &= 34 \\ \tau_{m2} &= 76\,. \end{aligned}$$

For the dynamics of the observer reference system $\boldsymbol{A}_S$ the following values are chosen to achieve a fast convergence of the observation error:

$$\begin{gathered} a_1 = 0{,}001\,, \quad a_2 = 0{,}002\,, \quad a_3 = 0{,}003\,, \\ \Rightarrow h_0 = -6 \cdot 10^{-9}\,, \quad h_1 = -1{,}1 \cdot 10^{-4}\,, \quad h_3 = -6 \cdot 10^{-3}\,. \end{gathered} \tag{4.32}$$

If for even k the measurement time $\tau_m(k) = \tau_{m1} = 34$ and for odd k a $\tau_m(k) = \tau_{m2} = 76$ is chosen,

$$\boldsymbol{l}(k) = \boldsymbol{l}_1 = \begin{pmatrix} 1{,}934 \\ -0{,}012 \\ -0{,}03 \end{pmatrix}$$

is obtained for even k and

$$\boldsymbol{l}(k) = \boldsymbol{l}_2 = \begin{pmatrix} 0{,}94 \\ 0{,}012 \\ 1{,}527 \end{pmatrix}$$

for odd k. As a result of the design procedure, the corresponding dynamics of the observation error for even k

$$\boldsymbol{H}(k) = \boldsymbol{H}_1 = \begin{pmatrix} -0{,}934 & -65{,}762 & -1{,}934 \\ 0{,}012 & 1{,}407 & 0{,}012 \\ 1{,}03 & 121{,}019 & 1{,}03 \end{pmatrix}$$

and for odd k

$$\boldsymbol{H}(k) = \boldsymbol{H}_1 = \begin{pmatrix} 1{,}94 & 71{,}452 & 0{,}94 \\ -0{,}012 & 0{,}09 & -0{,}012 \\ -0{,}527 & 3{,}951 & -0{,}527 \end{pmatrix}$$

are obtained by applying Equation (4.25). The simulation of the observation error $\tilde{\boldsymbol{x}}(k) = \boldsymbol{x}(k) - \hat{\boldsymbol{x}}(k)$ and $\tilde{\boldsymbol{x}}_B(k) = \boldsymbol{x}_B(k) - \hat{\boldsymbol{x}}_B(k)$ using the matrices $\boldsymbol{H}_1$ and $\boldsymbol{H}_2$ with an initial error

$$\tilde{\boldsymbol{x}}(0) = \begin{pmatrix} 1 \\ 1 \\ 1 \end{pmatrix}$$

and

$$\tilde{\boldsymbol{x}}_B(0) = \begin{pmatrix} 43 \\ 163 \\ 36 \end{pmatrix}$$

is shown in Figure 4.8. The plots show that the observation error converges very fast. The simulation yields the observation errors

$$\tilde{\boldsymbol{x}}(4) = \begin{pmatrix} 0{,}0021 \\ -0{,}00001 \\ -0{,}00057 \end{pmatrix}$$

and

$$\tilde{\boldsymbol{x}}_B(4) = \begin{pmatrix} 1{,}572 \cdot 10^{-9} \\ -2{,}6 \cdot 10^{-6} \\ 0{,}0011 \end{pmatrix}$$

after four steps of k.

Remark. Considering a stationary operation mode of the SMB and no measurement noise, the proposed wave front observer has the property that the wave front states are determined exactly for $k \to \infty$ independently of an initial state $\hat{\boldsymbol{x}}(0)$. This property is due to the assumption that all separation columns have the same adsorption characteristic and all inter–column volumes are the same. A vanishing observer error means that the measurement error $e(k) = y(k) - \hat{y}(k)$ also vanishes and, hence, the wave front observer provides the *exact* values for the concentration measurements at $\tau = \tau_m(k)$ for the position z_m, if an alternating measurement time is applied. Therefore, in the stationary state,

$$\begin{aligned} \hat{c}(\tau_{m1}, z_m, k) &= c(\tau_{m1}, z_m, k) \\ \hat{c}(\tau_{m2}, z_m, k) &= c(\tau_{m2}, z_m, k) \end{aligned} \tag{4.33}$$

holds. This property is independent of the setup $(n_{c,I}/n_{c,II}/n_{c,III}/n_{c,IV})$ or the operation point of the SMB and holds for all separation problems, to which the SMB is applied. The consequence of this property is that due to the slow dynamics of the SMB compared to that of the wave front observer, the wave front reconstruction based on the *autonomous* linear wave front model is a suitable means to track the transient of the SMB (see Remark below).

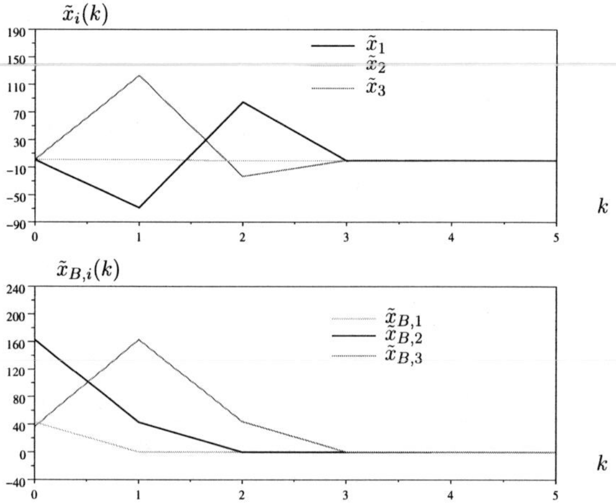

Figure 4.8: Observation error simulation for the original system and the system in observer canonical form

Solution of the wave front observation problem. Considering that only one concentration measurement is available per switching period k, the time–varying nature of the wave front state–space model, which is given by

$$\begin{aligned} \boldsymbol{x}(k+1) &= \underbrace{\begin{pmatrix} 1 & 0 & 0 \\ 0 & 1 & 0 \\ L & T_S & 1 \end{pmatrix}}_{\boldsymbol{A}} \boldsymbol{x}(k) \\ y(k) &= \underbrace{\begin{pmatrix} z_m & \tau_m(k) & 1 \end{pmatrix}}_{\mathbf{c}'(k)} \boldsymbol{x}(k)\,, \end{aligned} \tag{4.34}$$

has to be regarded. A Luenberger observer for the LTV system was applied to solve the observation problem:

$$\begin{aligned} \hat{\boldsymbol{x}}(k+1) &= \boldsymbol{A}\,\hat{\boldsymbol{x}}(k) + \boldsymbol{l}(k)\,e(k) \\ \hat{y}(k) &= \mathbf{c}'(k)\,\hat{\boldsymbol{x}}(k) \\ e(k) &= y(k) - \hat{y}(k)\,. \end{aligned}$$

The wave front is determined by

$$\hat{c}(\tau, z, k) = \mathrm{e}^{\begin{pmatrix} z & \tau & 1 \end{pmatrix} \hat{\boldsymbol{x}}(k)}\,.$$

The observer gain $\boldsymbol{l}(k)$ was determined based on the choice of the dynamics of the wave front observation error in observer canonical form considering an alternating measurement time with

$$\tau_m(k) \in \{\tau_{m1}, \tau_{m2}\}\,,\ \tau_m(k) \neq \tau_m(k-1)\,.$$

Therefore, two different gain vectors $\boldsymbol{l}_1$ and $\boldsymbol{l}_2$ are obtained.

Solution 4.3.1 *The solution of Task 4.3.1 is given by the observer*

$$\begin{cases} \hat{\boldsymbol{x}}(k+1) = \boldsymbol{A}\,\hat{\boldsymbol{x}}(k) + \boldsymbol{l}_1\,e(k) & \textit{for}\, k = 0, 2, 4, 6, \ldots \\ \hat{\boldsymbol{x}}(k+1) = \boldsymbol{A}\,\hat{\boldsymbol{x}}(k) + \boldsymbol{l}_2\,e(k) & \textit{for}\, k = 1, 3, 5, 7, \ldots \end{cases}\,. \tag{4.35}$$

□

The wave front observer which is obtained from this derivation corresponds to the observer represented in Figure 4.6. The algorithm of the wave front observer consists of the following six steps:

1.	Wave front concentration measurement:	$c(\tau_m(k), z_m, k)$
2.	Calculation of the process output:	$y(k) = \ln(c(\tau_m(k), z_m, k))$
3.	Calculation of the observer output:	$\hat{y}(k) = \begin{pmatrix} z_m & \tau_m(k) & 1 \end{pmatrix} \hat{\boldsymbol{x}}(k)$
4.	Calculation of the observer error:	$e(k) = y(k) - \hat{y}(k)$
5.	Calculation of the observer state update:	$\hat{\boldsymbol{x}}(k+1) = \boldsymbol{A}\,\hat{\boldsymbol{x}}(k) + \boldsymbol{l}(k)\,e(k)$
6.	Provide the wave front:	$\hat{c}(\tau, z, k) = \mathrm{e}^{\begin{pmatrix} z & \tau & 1 \end{pmatrix}\hat{\boldsymbol{x}}(k)}$

Remark. As was shown in (Kleinert and Lunze, 2004), the wave front observer shows a good tracking behaviour in case of a variation of the continuous control or disturbance input of the SMB process and therefore applies well for the SMB control. The property of a vanishing observer output error $e(k) = 0|_{k\to\infty}$ in the stationary state of the SMB is not affected by a change of the SMB process parameters. Hence, an extension of the wave front model to consider the effect of the input $\boldsymbol{u}$ onto the wave front is not necessary for the solution of Task 4.1.1.

Solution 4.3.2 *A solution to the Task 4.1.1 is given by the wave front observer shown in Figure 4.6, for which an alternating measurement time $\tau_m(k)$ according to Equation (4.21) is considered.* □

4.4 Observation error analysis

4.4.1 General concepts

The wave front model presented in Section 4.2 provides an approximate description of the wave fronts of a real SMB plant. Therefore, the observer wave front $\hat{c}(\tau, z, k)$ is not an exact representation of the corresponding SMB wave front. In the following, the deviation between the observer and the SMB wave front is referred to as the observation error. In addition to the observation error, which is caused by a model mismatch, an error can occur due to measurement uncertainties. The impact of both, the modelling and the measurement error on the observation error is investigated in the following sections.

The underlying goal of the wave front observation is to reconstruct the SMB wave fronts to be able to determine the recycling and the product stream impurity concentrations, and to recognise disturbances. To fulfil these requirements, the representation of the SMB wave front c_i by the observer wave front $\hat{c}_i$ as well as the wave front concentrations $\hat{c}_{max,i}$ provided by the wave front observer have to show sufficient accuracy in the stationary state, in case of a dynamic transition and in case of measurement errors.

In addition to the choice of the error dynamics the necessary parameters for the design of the wave front observer are the measurement times $\tau_{m1,i}$ and $\tau_{m2,i}$. The aim of the observation error analysis is to derive a guideline for the choice of the measurement times such that the desired accuracy of the observation result is met.

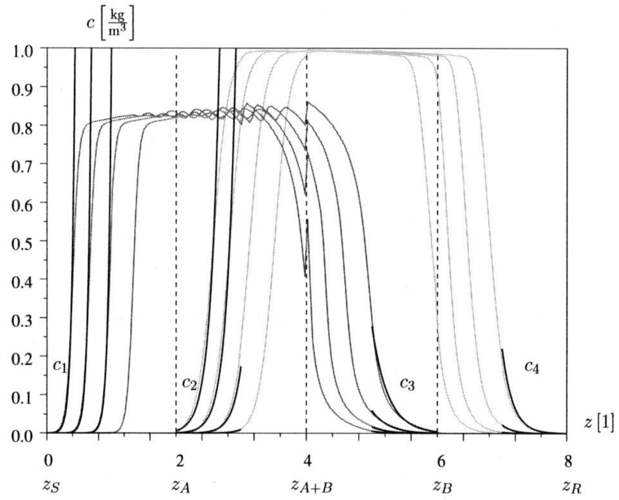

Figure 4.9: Stationary state concentration profiles (gray) and observer wave fronts (black) of a Tocopherol separation

The observation error analysis is performed based on the example SMB separation of Tocopherol shown in Figure 4.9. The figure shows the stationary concentration profiles and the observer wave fronts during one switching period. Because of the Anti–Langmuir–like adsorption, the desorption wave fronts c_1 and c_2 are steeper than the adsorption wave fronts c_3 and c_4.

The analysis of the observation error is performed as follows: In Section 4.4.2, the concept of the observation error and the mean observation error is introduced. In Section 4.4.3 the stationary observation error is analysed for the example process and a guideline for the choice of measurement times is given. Section 4.4.4 presents the analysis of measurement error impacts on the observation error for the stationary state. The tracking behaviour in case of a dynamical transition of the SMB is not considered here. It is addressed in (Kleinert and Lunze, 2004).

4.4.2 Observation error and mean observation error

The wave front observation error analysis refers to a model mismatch analysis and validation of a distributed–parameter system. Two approaches are considered here. The first approach is based on the analysis of the observation error for a given time point τ' and a given position z'. The second approach investigates the mean observation error over a given time horizon $[\tau_1, \tau_2]$ at a given position z'. Both approaches are used to validate the observation error. The approaches are introduced in the following paragraphs.

Observation error. Assuming a given wave front $c_i(\tau, z, k)$, $i = 1, 2, 3, 4$ of a real SMB process in the stationary operation mode and the corresponding stationary state observer wave front $\hat{c}_i(\tau, z, k)$, the *observation error* $\delta c_i(\tau, z, k)$ is determined by

$$\delta c_i(\tau', z', k) = c_i(\tau', z', k) - \hat{c}_i(\tau', z', k) \tag{4.36}$$

for a specified time τ' and position z'. Concerning the control of SMB processes, the observation error with respect to the largest impurity concentrations in the recycling and product streams is of interest. For the desorption wave fronts with $\tau' = T_S$ and $z' = z_{m,i} = 1$ this observation error is

$$\boxed{\delta c_i(T_S, z_{m,i}, k) = c_i(T_S, 1, k) - \hat{c}_i(T_S, 1, k)\,, \text{ for } i = 1, 2} \tag{4.37}$$

and for the adsorption wave fronts with $\tau' = 0$ and $z' = z_{m,i} = 0$ it is

$$\boxed{\delta c_i(0, z_{m,i}, k) = c_i(0, 0, k) - \hat{c}_i(0, 0, k)\,, \text{ for } i = 3, 4\,.} \tag{4.38}$$

Mean observation error. The impact of the continuous control and disturbance input can only be tracked by the wave front observer, if the observer wave front is a good approximation of the corresponding SMB wave front. Because the shape and the propagation of the SMB wave fronts are almost constant during one switching period (see Assumption 4.2.2), the propagation of $c(\tau, z, k)$ (or $\hat{c}(\tau, z, k)$) over $z \in [0, L]$ and $\tau \in [0, T_S]$ can be transformed to the evolution of the wave front at a given position z' over $\tau \in [0, T_S]$ and vice versa. This is shown in Figure 4.10 and 4.11 for the desorption wave front c_2 and the adsorption wave front c_3 of the example shown in Figure 4.9, together with the corresponding observer wave front. The left plot of the figures shows the wave fronts in z during one switching period for $\tau = 0$, $\tau = \tau_{m1}$, $\tau = \tau_{m2}$ and $\tau = T_S$, and the right plot shows the trajectory of the wave fronts at the respective measurement position z_m over $\tau \in [0, T_S]$ (the marks show the coincidence of the SMB and the observer wave fronts at the measurement times τ_{m1} and τ_{m2}). The absolute value of c_2 for $\tau = T_S$ and $z = z_m$ is $2{,}5 \cdot 10^{-3}\,\frac{\mathrm{kg}}{\mathrm{m}^3}$. c_3 takes the value $3{,}5 \cdot 10^{-3}\,\frac{\mathrm{kg}}{\mathrm{m}^3}$ for $\tau = 0$ and $z = z_m$.

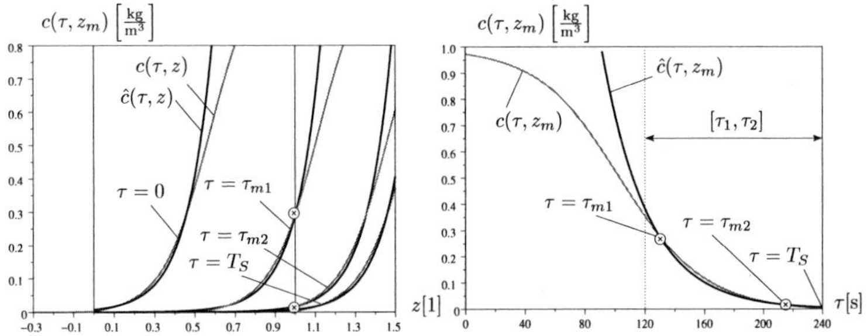

Figure 4.10: SMB (gray) and observer (black) desorption wave front in the stationary state

Figure 4.12 shows a close-up of the right plot of Figure 4.11. It shows that the observation error $\delta c(0, z_m)$ can take negative values.

Based on the idea of the transformation of the wave front trajectory, the mean observation error $\delta\bar{c}_i(z, k)$, $i = 1, 2, 3, 4$ is introduced, which describes the observation error $\delta c_i(\tau, z, k)$ over a given time horizon $\tau \in [\tau_1, \tau_2]$, $\tau_1, \tau_2 \in [0, T_S]$, at a specified position $z = z'$. The mean observation error is determined by

$$\delta\bar{c}_i(z', k) = \frac{1}{\tau_2 - \tau_1} \int\limits_{\tau_1}^{\tau_2} |\delta c_i(\tau, z, k)|\, d\tau \,.$$

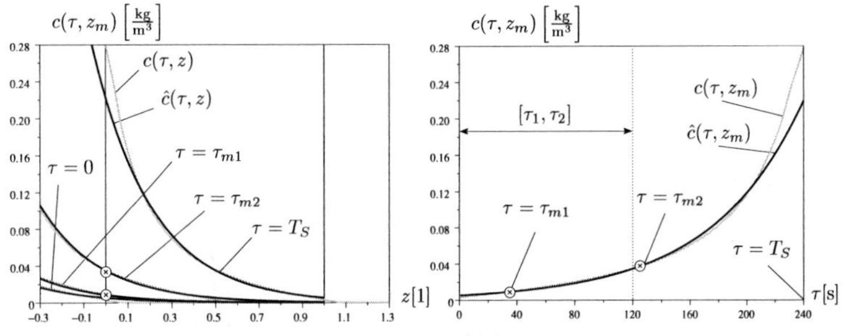

Figure 4.11: SMB (gray) and observer (black) adsorption wave front in the stationary state

Because the observer wave fronts shall represent those parts of the concentration profiles c_A and c_B which have low concentration values, the analysis range $[\tau_1, \tau_2]$ is chosen as shown in the right plots of Figure 4.10 and 4.11 for the desorption and the adsorption wave fronts:

$$\begin{aligned} \tau_1 &= \tfrac{1}{2} T_S\,, & \tau_2 &= T_S && \text{for the desorption wave front} \\ \tau_1 &= 0\,, & \tau_2 &= \tfrac{1}{2} T_S && \text{for the desorption wave front}\,. \end{aligned}$$

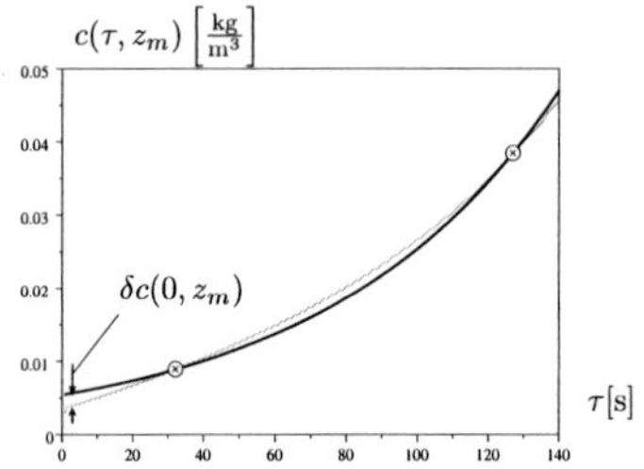

Figure 4.12: Close–up of the SMB (gray) and observer (black) adsorption wave front

For both types of wave fronts, the position $z' = z_m$ is chosen for the analysis. Hence, the mean observation error for the desorption wave fronts is determined by

$$\boxed{\delta\bar{c}_i(z_m, k) = \frac{2}{T_S} \int\limits_{\frac{1}{2}T_S}^{T_S} |\delta c_i(\tau, 1, k)|\, d\tau\,, \quad \text{for } i = 1, 2\,,} \tag{4.39}$$

and for the adsorption wave fronts by

$$\boxed{\delta\bar{c}_i(z_m, k) = \frac{2}{T_S} \int\limits_{0}^{\frac{1}{2}T_S} |\delta c_i(\tau, 0, k)|\, d\tau\,, \quad \text{for } i = 3, 4\,.} \tag{4.40}$$

4.4.3 Stationary observation error

In this section, the impact of the choice of the measurement times τ_{m1} and τ_{m2} of the wave front observers on the observation error in the stationary state is investigated. An analytic expression for the observation error is derived, for which the indicator $i = 1, 2, 3, 4$ of the wave fronts is omitted because it applies to any of the four wave fronts.

The investigation of the observation error for various measurement times yields a proposition for the choice of the τ_{m1} and τ_{m2} for the desorption and the adsorption wave fronts. Because the stationary state is considered, it is not necessary to consider the dependence of the respective variables onto the period k. Therefore, the analysis is performed for one switching period and the variable k is omitted.

Analytic expression for the observation error. In the stationary state, the observer wave front concentrations coincide with the measured SMB wave front concentrations c_{m1} and c_{m2} at the measurement times τ_{m1}, τ_{m2} and the position z_m (see Equation (4.33) and e.g. Figure 4.12):

$$\begin{aligned} \hat{c}(\tau_{m1}, z_m) &= c(\tau_{m1}, z_m) = c_{m1} \\ \hat{c}(\tau_{m2}, z_m) &= c(\tau_{m2}, z_m) = c_{m2} \,. \end{aligned} \tag{4.41}$$

The observer wave front is determined by

$$\hat{c}(\tau, z, k) = \mathrm{e}^{\hat{x}_1 z + \hat{x}_2 \tau + \hat{x}_3} \,. \tag{4.42}$$

With Equation (4.41) and (4.42) the stationary wave front states are determined by

$$\hat{x}_1 = T_S \frac{\ln(c_{m1}) - \ln(c_{m2})}{\tau_{m2} - \tau_{m1}} \,, \tag{4.43}$$

$$\hat{x}_2 = -\frac{1}{T_S}\,\hat{x}_1 = -\frac{\ln(c_{m1}) - \ln(c_{m2})}{\tau_{m2} - \tau_{m1}} \,, \tag{4.44}$$

$$\begin{aligned} \hat{x}_3 &= \ln(c_{m1}) - \hat{x}_1 \left(z_m - \frac{\tau_{m1}}{T_S} \right) \\ &= \frac{\ln(c_{m1}) \cdot (\tau_{m1} - T_S\, z_m) - \ln(c_{m2}) \cdot (\tau_{m2} - T_S\, z_m)}{\tau_{m2} - \tau_{m1}} \,, \end{aligned} \tag{4.45}$$

where the factor $-\frac{1}{T_S}$ in Equation (4.44) denotes the stationary propagation velocity of the wave front. Applying Equations (4.42), (4.43), (4.44) and (4.45) to Equation (4.36) yields

$$\boxed{\delta c(\tau', z') = c(\tau', z') - \underbrace{\frac{c_{m1}^{\left(\frac{(z'-z_m)\,T_S + \tau_{m2} - \tau'}{\tau_{m2} - \tau_{m1}}\right)}}{c_{m2}^{\left(\frac{(z'-z_m)\,T_S + \tau_{m1} - \tau'}{\tau_{m2} - \tau_{m1}}\right)}}}_{\hat{c}(\tau', z')}} \tag{4.46}$$

as an explicit description of the observation error $\delta c(\tau, z)$ in the stationary operation mode of the SMB process, where $\tau' \in [0, T_S]$ and $z' \in [0, L]$ holds. Applying Equation (4.46) to Equation (4.37) or (4.38), respectively, allows for the quantitative specification of the stationary observation error for a given SMB process.

Impact of the choice of the measurement times. The analysis of the observation error is based on the example process shown in Figure 4.9: For the analysis, δc and $\delta \bar{c}$ are determined for a variation of the measurement times τ_{m1} and τ_{m2}, regarding $\tau_{m1} \overset{!}{<} \tau_{m2}$, for the wave fronts c_2 and c_3. The observation errors are determined according to Equation (4.37), (4.38), (4.39) and (4.40). Different results and conclusions are obtained for the desorption and for the adsorption wave fronts.

First, the desorption wave front c_2 is considered: The observation errors δc and $\delta \bar{c}$ are determined for a variation of the measurement times in the range of

$$\begin{aligned} \tau_{m1} &\in \left[\tfrac{1}{2} T_S, T_S\right] = [120, 240] \\ \tau_{m2} &\in \left[\tfrac{1}{2} T_S, T_S\right] = [120, 240] \,. \end{aligned}$$

The result is shown in Figure 4.13. The plots show the surface of the observation errors for the considered range of τ_{m1} and τ_{m2}.

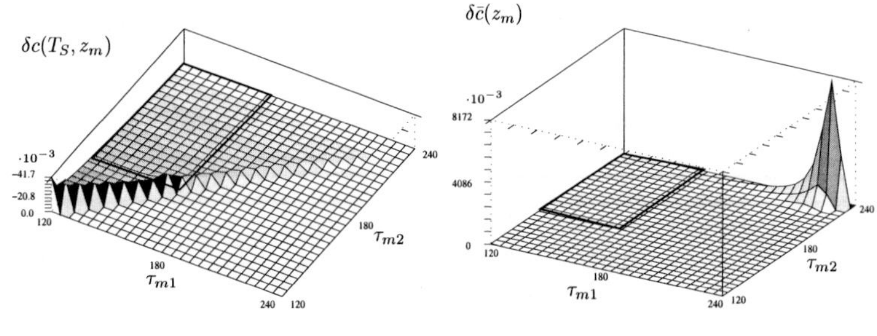

Figure 4.13: Stationary observation error for the desorption wave front c_2 in $\left[\frac{\text{kg}}{\text{m}^3}\right]$

It can be seen from the left plot that the observation error δc becomes zero, if $\tau_{m2} \to T_S$, which can also be derived by applying Equation (4.37) to Equation (4.41) for $\tau_{m2} = T_S$. δc increases if τ_{m2} is reduced. A further increase occurs if the measurement times are close to each other. The largest error occurs if both measurement times are small. Considering the right surface plot it seems that only for $\tau_{m1}, \tau_{m2} \to T_S$ the mean observation error $\delta \bar{c}$ becomes very large, which has to be avoided. To be able to analyse the observation error in more detail, a new range for τ_{m1} and τ_{m2} is chosen, which corresponds to the marked areas in the plots of Figure 4.13:

$$\begin{aligned} \tau_{m1} &\in [\tfrac{1}{2} T_S, \tfrac{2}{3} T_S] &= [120, 160] \\ \tau_{m2} &\in [\tfrac{2}{3} T_S, T_S] &= [160, 240]\,. \end{aligned}$$

The results for this range are shown in Figure 4.14.

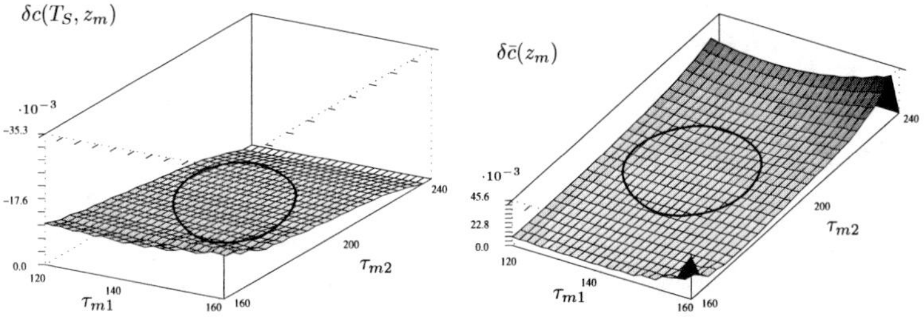

Figure 4.14: Stationary observation error for the desorption wave front c_2 for a new range of τ_{m1}, τ_{m2} in $\left[\frac{\text{kg}}{\text{m}^3}\right]$

For the observation error, the same result as for the previous analysis is obtained. δc becomes zero, if $\tau_{m2} = T_S$. It increases for a reduction of τ_{m2}. The variation of τ_{m1} for the considered range has a marginal influence. For the mean observation error $\delta\bar{c}$ an increase occurs, if $\tau_{m2} \to T_S$. There is only a marginal change for the variation of τ_{m1}. However, an increase occurs if τ_{m1} is on the limits of the considered interval.

For the choice of τ_{m1} and τ_{m2}, a compromise between the resulting observation error and mean observation error has to be made. For the present example it is concluded that for desorption wave fronts, for which a nonlinear Anti–Langmuir like adsorption occurs, the measurement times τ_{m1} and τ_{m2} shall be chosen from the intervals

$$\begin{aligned} \tau_{m1} &\in [0{,}5\,T_S, 0{,}65\,T_S] \\ \tau_{m2} &\in [0{,}7\,T_S, 0{,}9\,T_S]\,, \end{aligned} \tag{4.47}$$

which corresponds to

$$\begin{aligned} [0{,}5\,T_S, 0{,}65\,T_S] &\approx [120, 155] \\ [0{,}7\,T_S, 0{,}9\,T_S] &\approx [165, 215] \end{aligned}$$

for the considered example, and

$$\tau_{m2} \geq \tau_{m1} + 0{,}2\,T_S$$

(see the areas in Figure 4.14 marked by black lined ovals; for this area, $\delta c \approx 7 \cdot 10^{-3}\,\frac{\mathrm{kg}}{\mathrm{m}^3}$ and $\delta\bar{c} \approx 1{,}5 \cdot 10^{-2}\,\frac{\mathrm{kg}}{\mathrm{m}^3}$).

For the analysis of the adsorption wave front c_3 the measurement times are varied in the range of

$$\begin{aligned} \tau_{m1} &\in [0, \tfrac{2}{3}\,T_S] &= [0, 160] \\ \tau_{m2} &\in [0, \tfrac{2}{3}\,T_S] &= [0, 160]\,. \end{aligned}$$

The observation errors δc and $\delta\bar{c}$ are determined using Equation (4.38) and (4.40). The results are shown in Figure 4.15.

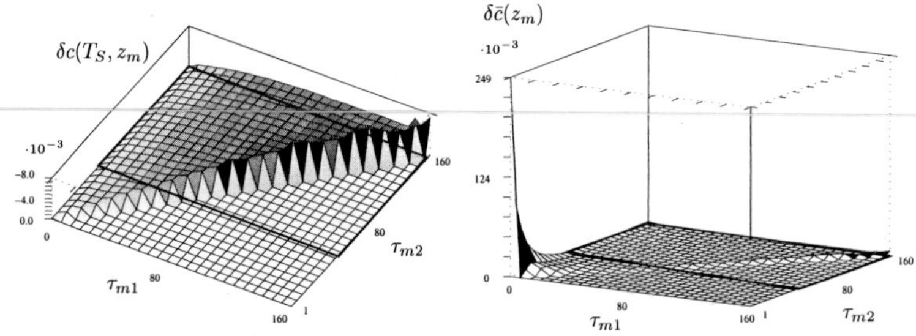

Figure 4.15: Stationary observation error for the adsorption wave front c_3 in $\left[\frac{\mathrm{kg}}{\mathrm{m}^3}\right]$

The left plot shows that the observation error δc becomes zero, if $\tau_{m1} \to 0$. This can also be derived by applying Equation (4.38) to Equation (4.41) for $\tau_{m1} = 0$. δc increases with increasing τ_{m1}. Furthermore, the error increases, if the the measurement times τ_{m1} and τ_{m2} are close to each other. The largest error occurs if both measurement times take values close to T_S. The mean observation error $\delta\bar{c}$ (right plot) becomes very large for $\tau_{m1}, \tau_{m2} \to 0$. It decreases for an increase of τ_{m2} while $\tau_{m1} \to 0$. Furthermore, an increase of $\delta\bar{c}$ can be observed for $\tau_{m1}, \tau_{m2} \to T_S$. To be able to analyse $\delta\bar{c}$ in more detail, a new range for τ_{m2} is considered while keeping the previously chosen range for τ_{m1}:

$$\begin{aligned} \tau_{m1} &\in [0, \tfrac{2}{3}\,T_S] &= [0, 160] \\ \tau_{m2} &\in [\tfrac{1}{6}\,T_S, \tfrac{2}{3}\,T_S] &= [40, 160]\,. \end{aligned}$$

The new range corresponds to the marked areas in Figure 4.15.

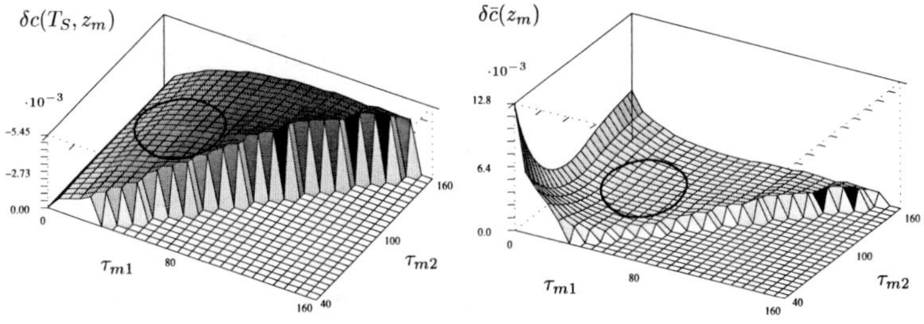

Figure 4.16: Stationary observation error for the adsorption wave front c_3 for a new range of τ_{m1}, τ_{m2} in $\left[\frac{\text{kg}}{\text{m}^3}\right]$

The result is shown in Figure 4.16. The left plot reveals the same result as for the previous analysis. The observation error δc becomes zero, if $\tau_{m1} \rightarrow 0$. It increases with an increase of τ_{m1}. It can be seen from the right plot that the mean observation error $\delta\bar{c}$ increases, if $\tau_{m1} \rightarrow 0$ or $\tau_{m1} \rightarrow \frac{2}{3}T_S$. $\delta\bar{c}$ increases for low and for large τ_{m2}. In the intermediate range, $\delta\bar{c}$ changes only moderately. A good compromise for small values of δc and $\delta\bar{c}$ is obtained for

$$\begin{aligned} \tau_{m1} &\in [0{,}1\,T_S, 0{,}35\,T_S] \\ \tau_{m2} &\in [0{,}3\,T_S, 0{,}6\,T_S]\,, \end{aligned} \tag{4.48}$$

which corresponds to

$$\begin{aligned} [0{,}1\,T_S, 0{,}35\,T_S] &\approx [25, 85] \\ [0{,}3\,T_S, 0{,}6\,T_S] &\approx [75, 145] \end{aligned}$$

for the considered example (see the oval marked areas in the surface plots of Figure 4.16, for which $\delta c \approx 4 \cdot 10^{-3}$ and $\delta\bar{c} \approx 1{,}3 \cdot 10^{-3}$). Note that for this example the wave front underlies mainly the linear adsorption. This is the reason for the good fitting of the SMB wave front by the observer wave front, also if the measurement times are close to each other. If, however, nonlinear Langmuir–like adsorption has to be considered, the upper bound for τ_{m1} and τ_{m2} should be reduced for the adsorption wave front.

Conclusion. Based on the previous analysis of the observation error the following suggestion is made for the choice of the measurement times: *Nonlinear adsorption* causes larger observation errors. In this case, the measurement times for the *desorption wave fronts* should be chosen from the intervals

$$\begin{aligned}\tau_{m1} &\in [0{,}5\,T_S, 0{,}65\,T_S]\\ \tau_{m2} &\in [0{,}7\,T_S, 0{,}9\,T_S]\,,\end{aligned}$$

as indicated in Equation (4.47). The difference between the measurement times should be about $0{,}2\,T_S$.

If *linear adsorption* occurs, the intervals can be enlarged for the *desorption wave fronts* to

$$\begin{aligned}\tau_{m1} &\in [0{,}4\,T_S, 0{,}7\,T_S]\\ \tau_{m2} &\in [0{,}65\,T_S, 0{,}9\,T_S]\,,\end{aligned}$$

which is obtained from interpreting the results given in Equation (4.48) for desorption wave fronts.

In case of *nonlinear adsorption*, the measurement times for the *adsorption wave fronts* should be chosen from the intervals

$$\begin{aligned}\tau_{m1} &\in [0{,}1\,T_S, 0{,}3\,T_S]\\ \tau_{m2} &\in [0{,}35\,T_S, 0{,}5\,T_S]\,.\end{aligned}$$

The result is obtained from interpreting Equation (4.47) for adsorption wave fronts. The difference between the measurement times should be about $0{,}2\,T_S$.

If *linear adsorption* occurs, the intervals can be chosen for the *adsorption wave fronts* to

$$\begin{aligned}\tau_{m1} &\in [0{,}1\,T_S, 0{,}35\,T_S]\\ \tau_{m2} &\in [0{,}3\,T_S, 0{,}6\,T_S]\,,\end{aligned}$$

according to Equation (4.48).

4.4.4 Impact of measurement errors

In this section the impact of measurement errors onto the observation error is investigated. The impact is analysed considering the wave front concentrations which occur at the corresponding measurement position z_m over the time interval $\tau \in [\tau_1, \tau_2]$. Two kinds of measurement errors are considered: On the one hand, it is supposed that a *constant temporal offset of the measurement time point* occurs. On the other hand, a *constant offset of the measured concentration value* is assumed. The considered types of errors are described in detail in the following paragraphs. The investigation of the error impact is performed for the wave fronts c_2 and c_3 of the example process shown in Figure 4.17 (Ibuprofen separation) and is discussed in the following. The stationary operation mode is considered. Therefore, the dependence of the parameters upon k is not indicated.

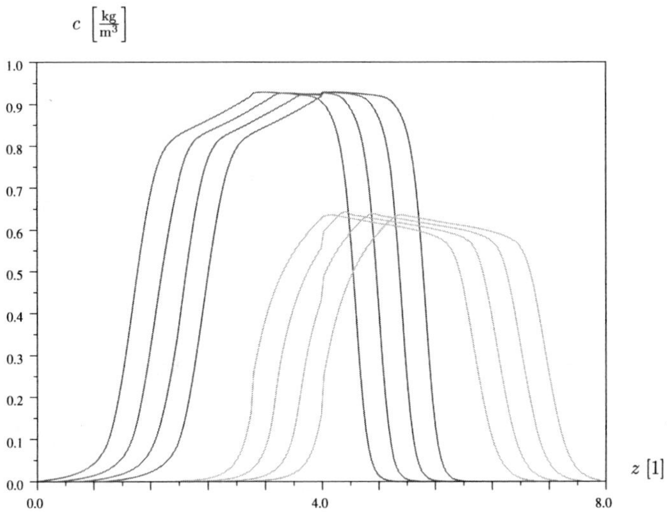

Figure 4.17: Concentration profile snapshots during one switching period

Measurement uncertainties. Two kinds of measurement uncertainties, which lead to the measurement errors, are considered. First, the time point at which the measurement is actually recorded, might differ from the specified measurement time τ_m by an amount of $\pm\Delta\tau_m$. Second, the concentration measurement might show a constant concentration offset of $\pm\Delta c_m$ with respect to the real concentration value. If both effects occur simultaneously, they can cancel out each other. Therefore, for the investigation of the impact of the measurement error, the following combination of both effects leading to the largest measurement error are considered for the desorption and the adsorption wave fronts:

$$\text{Desorption wave fronts:} \quad (\Delta\tau_m < 0, \Delta c_m > 0)$$
$$(\Delta\tau_m > 0, \Delta c_m < 0)\,,$$

$$\text{Adsorption wave fronts:} \quad (\Delta\tau_m > 0, \Delta c_m > 0)$$
$$(\Delta\tau_m < 0, \Delta c_m < 0)\ .$$

Figures 4.18 and 4.19 show, how the measurement uncertainties influence the actual measurement value c_m.

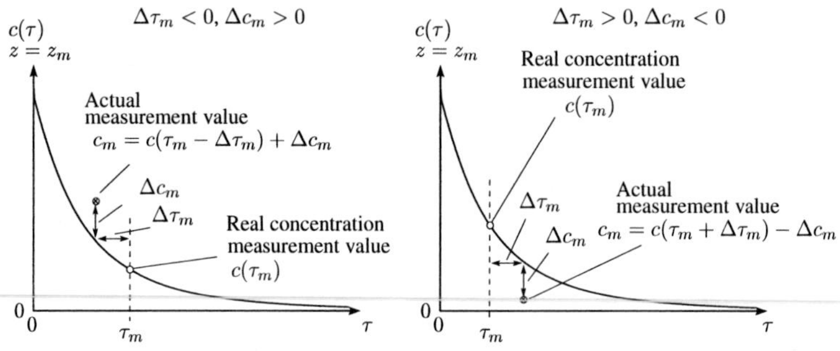

Figure 4.18: Measurement error for desorption wave fronts

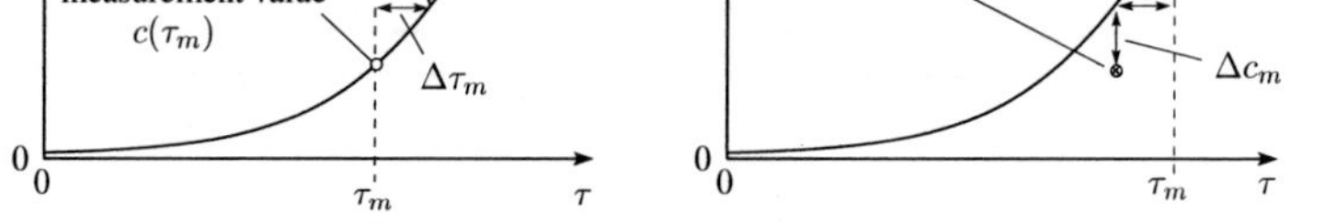

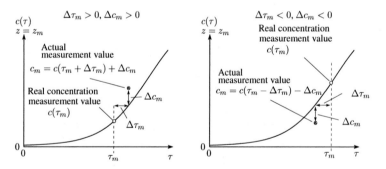

Figure 4.19: Measurement error for adsorption wave fronts

Measurement error. The actual measurement value

$$c_m = c(\tau_m \pm \Delta\tau_m) \pm \Delta c_m$$

is that value which is provided by the measurement unit to the wave front observer as the concentration measurement at the measurement time τ_m. In other words, c_m is assigned to the measurement time τ_m by the observer regardless of the correctness of the numerical value of c_m. Considering the aforementioned combinations of the measurement uncertainties for each measurement time, the lumped measurement error dc_m with

$$\begin{aligned} dc_m(\tau_m, z_m) &= c(\tau_m, z_m) - c_m \\ &= c(\tau_m, z_m) - (c(\tau_m \pm \Delta\tau_m) \pm \Delta c_m) \end{aligned}$$

is obtained. Because the real concentration can only be positive or zero it is assumed that the minimum value of the considered observer wave front is zero, if a negative measurement value is provided.

Analysis of the observation error considering measurement errors. No information about the measurement errors is provided to the wave front observer and, hence, the actual measurement $c_m = c(\tau_m \pm \Delta\tau_m) \pm \Delta c_m$ is considered as the actual concentration of the SMB wave front at the measurement time τ_m and the measurement position z_m. In the stationary state, the model wave front $\hat{c}(\tau, z)$, which is provided by the observer, coincides with the measurements at the measurement times τ_m (and at the measurement position z_m), i.e.

$$\hat{c}(\tau_m, z_m) = c_m .$$

Because of the coincidence of the observer wave front and the measurement at the measurement time point, the observation error

$$\begin{aligned} \delta c_m(\tau_m, z_m) &= c(\tau_m, z_m) - \hat{c}_m(\tau_m, z_m) \\ &= c(\tau_m, z_m) - c_m \\ &= c(\tau_m, z_m) - (c(\tau_m \pm \Delta\tau_m) \pm \Delta c_m) \\ &= dc_m \end{aligned}$$

occurs, which is equal to the measurement error dc_m.

If the temporal offset $\Delta\tau_m$ and the concentration offset Δc_m are within a given range

$$\begin{aligned} -\Delta\tau_{m,max} &\leq \Delta\tau_m \leq \Delta\tau_{m,max} \\ -\Delta c_{m,max} &\leq \Delta c_m \leq \Delta c_{m,max} , \end{aligned}$$

but are time–invariant, the measurement error dc_m will either be larger than zero or lower than zero for both measurement times τ_{m1} and τ_{m2}. This means that the wave front provided by the observer lies within a bound limited by two curves, which each cross the upper or the lower

Desorption wave front c_2	$\tau_{m1} = 40$ s	$\tau_{m2} = 90$ s
c_m	$26{,}3 \cdot 10^{-3}$	$3{,}9 \cdot 10^{-3}$
dc_m	$+7{,}1 \cdot 10^{-3}$	$+5{,}3 \cdot 10^{-3}$
	$-6{,}1 \cdot 10^{-3}$	$-3{,}8 \cdot 10^{-3}$
Adsorption wave front c_3	$\tau_{m1} = 30$ s	$\tau_{m2} = 90$ s
c_m	$23{,}2 \cdot 10^{-3}$	$154{,}5 \cdot 10^{-3}$
dc_m	$+7{,}7 \cdot 10^{-3}$	$+19{,}9 \cdot 10^{-3}$
	$-9{,}1 \cdot 10^{-3}$	$-28{,}6 \cdot 10^{-3}$

Table 4.1: Measurement value and measurement errors for the example wave fronts c_2 and c_3 in $\left[\frac{\text{kg}}{\text{m}^3}\right]$

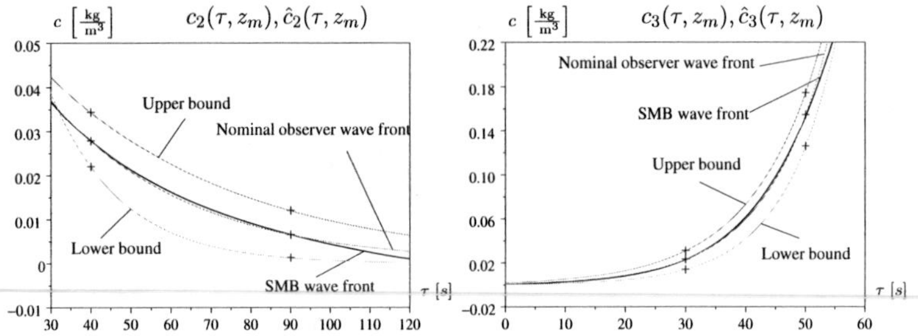

Figure 4.20: Nominal model wave front and bounds for the model wave fronts for constant measurement errors

bounds of the measurement errors. The observer wave front of the nominal case, for which no measurement error is considered, also lies within this bound of the wave fronts. Figure 4.20 shows the nominal observer wave front, and the upper and lower bounds for the example wave fronts for the range of $\Delta\tau_m \leq 1{,}5$ s and $\Delta c_m \leq 5 \cdot 10^{-3} \frac{\text{kg}}{\text{m}^3}$. Table 4.1 shows the absolute values of dc_m for the considered example wave fronts and the respective measurement times.

If it is assumed that $\Delta\tau_m$ and Δc_m lie within given ranges but take alternate values for each measurement time τ_{m1} and τ_{m2}, then two more cases providing upper and lower bounds for the model wave fronts occur. Figure 4.21 shows for each of the four SMB wave fronts the upper and lower bound for the model wave fronts and the two additional cases for $\Delta\tau_m$ and Δc_m within the given ranges. Case 1 assumes the combination of the upper measurement bound of the first measurement at τ_{m1} with the lower measurement bound of the second measurement at τ_{m2}. Case 2 assumes the combination of the lower measurement bound of the first measurement with the upper measurement bound of the second measurement.

To validate the impact of the measurement errors, the observation error δc and the mean observation error $\delta\bar{c}$ are determined for each of the cases shown in Figure 4.20 and 4.21. As proposed in Section 4.4.2 the mean observation error of the desorption wave front is determined for the time

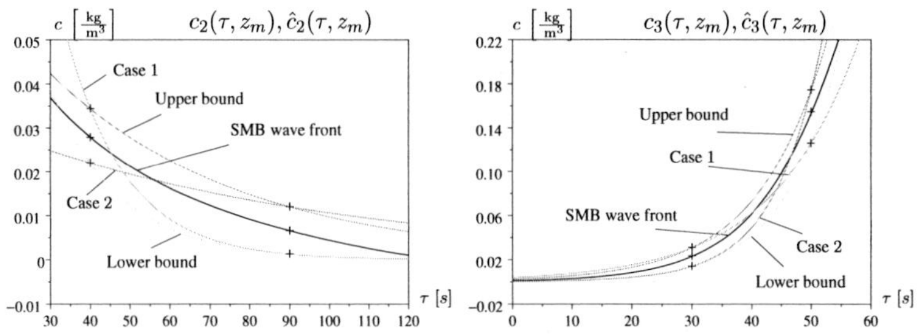

Figure 4.21: Bounds for the model wave fronts for alternating measurement errors

	Upper bound	Case 1	Case 2	Lower bound	Nominal case
$\delta c_m(T_S, z_m)$	$-5{,}33 \cdot 10^{-3}$	$8{,}75 \cdot 10^{-4}$	$-7{,}29 \cdot 10^{-3}$	$8{,}21 \cdot 10^{-4}$	$-1{,}69 \cdot 10^{-3}$
$\delta \bar{c}_m(z_m)$	$5{,}6 \cdot 10^{-3}$	$4{,}98 \cdot 10^{-3}$	$5{,}04 \cdot 10^{-3}$	$5{,}23 \cdot 10^{-3}$	$6{,}58 \cdot 10^{-4}$

Table 4.2: Observation error for the wave front c_2 in $\left[\frac{kg}{m^3}\right]$

interval $\tau \in [\frac{1}{2} T_S, T_S]$ at the position z_m. The mean observation error of the adsorption wave fronts is determined for the time interval $\tau \in [0, \frac{1}{2} T_S]$ at z_m.

Tables 4.2 and 4.3 show the observation error δc and the mean observation error $\delta \bar{c}$ for each considered case of measurement errors and for the nominal case.

	Lower bound	Case 1	Case 2	Upper bound	Nominal case
$\delta c_m(T_S, z_m)$	$-1{,}43 \cdot 10^{-3}$	$-2{,}89 \cdot 10^{-3}$	$5{,}54 \cdot 10^{-4}$	$3{,}52 \cdot 10^{-4}$	$-4{,}71 \cdot 10^{-4}$
$\delta \bar{c}_m(z_m)$	$1{,}39 \cdot 10^{-2}$	$1{,}36 \cdot 10^{-2}$	$2{,}49 \cdot 10^{-2}$	$1{,}14 \cdot 10^{-2}$	$5{,}29 \cdot 10^{-3}$

Table 4.3: Observation error for the wave front c_3 in $\left[\frac{kg}{m^3}\right]$

Discussion. The values for the observation error δc and the mean observation error $\delta \bar{c}$ show that for each wave front the nominal case is in a good agreement with the SMB wave front. Almost all other cases show moderate observation errors of about $5 \cdot 10^{-3} \frac{kg}{m^3}$ for the desorption wave front and about $1{,}5 \cdot 10^{-2} \frac{kg}{m^3}$ for the adsorption wave front.

With respect to the considered measurement error dc_m (Table 4.1) the analysis of the measurement error impact shows that the values of δc and $\delta \bar{c}$ have values of the order of dc_m. Hence, the measurement error dc_m can be applied for an approximate determination of δc and $\delta \bar{c}$.

4.5 Summary

A reconstruction of the SMB wave fronts based on selected discrete–time concentration measurements and the derivation of an observation algorithm of low complexity was presented in this chapter. An explicit functional description of the shape and the propagation was used for the modelling to derive a simple wave front model including a third–order state–space model of the wave front parameter dynamics. The linear autonomous wave front model was obtained, which does not include parameters of the detailed plant model except the switching time T_S. The observer parameters are the observer error dynamics and the measurement times τ_{m1}, τ_{m2}, which have to be chosen to compute the observer gain $\boldsymbol{l}$.

In the stationary state, the observer has the property of exactness with respect to the wave front concentration measurements. The analysis of the observation error shows that the wave front observer is a suitable means for the reconstruction of the SMB wave fronts. A rule for the choice of the measurement times was derived.

The wave front observer was derived for and applied to SMB processes considering synchronous port switchings. Because all wave fronts are considered as independent systems $\mathcal{W}_i$ to which each one wave front observer is applied, it is principally possible to apply this concept of wave front reconstruction to the VARICOL process.

Chapter 5

Continuous process control

This Chapter addresses the control of SCC processes to which a time–invariant switching pattern is applied. The free operation parameters, which serve as control input, are the internal fluid flow rates. The controller for this setup is referred to as the continuous controller of SCC processes.
Because the dynamics of SCC processes with a time–invariant switching pattern has specific properties, decentralised discrete–time PI controllers can be applied. The control concept is based on discrete–time wave front concentration and purity measurements. Two controller schemes are investigated. The first considers the direct use of the concentration measurements as controlled plant output signals. The second uses the concentration measurements for the reconstruction of the controlled plant output by wave front observers.
The derivation of the controller design model, the controller design and numerical simulations are presented for three SCC processes considering different plant configurations and mixture separations.

5.1 Control task

For the continuous control of SCC processes, a time–invariant switching pattern is considered, i.e. the port switching sequence is predetermined. The remaining control input signals are the continuous–variable internal fluid flow rates $\dot{m}_j$, $j = I, II, III, IV$. The control aim according to Task 1.2.1 is to drive the SCC process to the desired set–point during the start–up of the plant. Furthermore, step disturbances of the feed inlet concentration and a drift of the adsorbent package porosity, which change the product purity values, shall be attenuated by the control.

According to the SCC operation mode proposed in Sections 2.3.3 and 2.4.3 the impurity of the recycling stream should be kept low such that the wave fronts c_1 and c_4 mainly propagate within the sections I and IV. Then, the wave fronts can be manipulated separately by the flow rates $\dot{m}_I$ and $\dot{m}_{IV}$ such that the set–points $c_{set,1}$ and $c_{set,4}$ for the recycling impurities are reached. In this case, the impurity of the product streams $\dot{m}_A$ and $\dot{m}_B$ are mainly determined by the wave fronts c_2 and c_3, and the purity values r_A and r_B can be adjusted by the manipulation of the flow rates

$\dot{m}_{II}$ and $\dot{m}_{III}$.

Because the switching pattern is time–invariant and, hence, the switching of each port is equidistant in time, the considered SCC processes can be regarded as discrete–time systems, which are sampled in the moment of the switching of the solvent inlet S. The control input signals, i.e. the internal fluid flow rates, are considered to be constant during the switching periods and, therefore, are represented by $\dot{m}_j(k)$, $j = I, II, III, IV$.

The purity values, which are recorded according to the concept proposed in Section 2.5, are considered as discrete–time output signals. For the derivation of the continuous controllers it is assumed that the purity values are available with a time–delay of one period k. The purity values are represented by $r_A(k)$ and $r_B(k)$. It is assumed that one concentration measurement $c_i(\tau_{m,i}(k), z_{m,i}, k)$, $i = 1, 2, 3, 4$ is available per wave front c_i and per switching period.

The feed inlet concentrations $c_{A+B,A}$ and $c_{A+B,B}$ and the package porosity ε are unmeasured disturbance input signals. It is assumed that the signals are constant during the switching periods and, therefore, are represented by $c_{A+B,*}(k)$ and $\varepsilon(k)$.

Figure 5.1 shows the block diagram of the SCC plant with the input and output signals as it is considered for the control task.

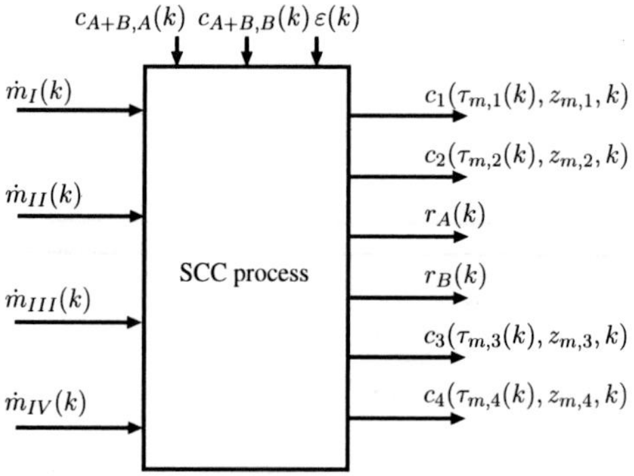

Figure 5.1: Input and output signals of SCC processes with a time–invariant switching pattern

The flow rates of the internal fluid flows are lumped to the control input vector

$$\boldsymbol{u}(k) = \begin{pmatrix} \dot{m}_I(k) \\ \dot{m}_{II}(k) \\ \dot{m}_{III}(k) \\ \dot{m}_{IV}(k) \end{pmatrix} .$$

The disturbance input variables are represented by

$$\boldsymbol{d}(k) = \begin{pmatrix} c_{A+B,A}(k) \\ c_{A+B,B}(k) \\ \varepsilon(k) \end{pmatrix} .$$

If a VARICOL process is considered, k has to be replaced by k_S.

For these SCC processes, the following control task is formulated in terms of Task 1.2.1:

Task 5.1.1 *Given is an SCC process with a time–invariant switching pattern and the input and output signals as shown in Figure 5.1. Manipulate the control input* $\mathbf{u}(k)$ *such that the purity values* $r_A(k)$ *and* $r_B(k)$ *as well as the wave front concentrations* $c_{m,1}(k) = c_1(\tau_{m,1}(k), z_{m,1}, k)$ *and* $c_{m,4}(k) = c_4(\tau_{m,4}(k), z_{m,4}, k)$ *reach a given set–point*

$$\boldsymbol{w} = \begin{pmatrix} c_{set,1} \\ r_{set,A} \\ r_{set,B} \\ c_{set,4} \end{pmatrix}$$

and are kept there after

- *the start–up of the plant,*
- *a step disturbance of* $c_{A+B,A}(k)$ *and* $c_{A+B,B}(k)$ *and*
- *a ramp disturbance of* $\varepsilon(k)$.

□

Two approaches to the solution of Task 5.1.1 are presented in the following sections, which refer to the continuous control concepts introduced in Section 1.3. The first approach considers the *measurement–based continuous control* which is realised without a reconstruction of the concentration profile (Section 5.2). The second approach refers to the *observation–based continuous control* (Section 5.3). The identification of the controller design model is discussed in Section 5.4. Section 5.5 presents application examples and simulation studies.

5.2 Measurement–based control

5.2.1 Plant model

For the measurement–based control, the following plant output vector is considered according to the control concept introduced in Section 1.3:

$$\boldsymbol{y}(k) = \begin{pmatrix} c_{m,1}(k) \\ r_A(k) \\ r_B(k) \\ c_{m,4}(k) \end{pmatrix} .$$

Two alternatives for the measurement of the wave front concentrations $c_{m,1}$ and $c_{m,4}$ can be taken into account. The first considers the largest impurity concentrations in the recycling stream as the measured wave front concentration:

$$\begin{aligned} c_{m,1}(k) &= c_{max,1}(z_S, k) &= c_1(0, z_S, k)\,, \\ c_{m,4}(k) &= c_{max,4}(z_S, k) &= c_4(T_S, z_S, k)\,. \end{aligned}$$

This concept requires the sampling of the recycling impurity shortly before and shortly after the switching of the solvent inlet port S at the position z_S. The second concept uses one sampling of the recycling impurity at $\tau = \frac{1}{2}T_S$:

$$\begin{aligned} c_{m,1}(k) &= c_1(\tfrac{1}{2}T_S, z_S, k)\,, \\ c_{m,4}(k) &= c_4(\tfrac{1}{2}T_S, z_S, k)\,. \end{aligned}$$

Figure 5.2 (a) shows the considered plant in form of a 4×4 multiple–input multiple–output (MIMO) system with four control input and four controlled output signals. As will be shown for the example processes, the dynamical input–output behaviour shows low cross couplings between the control input and the controlled output signals such that the system can be represented by 4 single–input single–output (SISO) systems (Figure 5.2 (b)). The dynamics of the four SISO systems can be represented by linear first order discrete–time systems (see the results of Section 5.5). For the dynamics of r_A and r_B the time delay by one switching period has to be regarded.

5.2.2 Controller design

The system structure allows for the application of decentralised discrete–time PI controllers (see e.g. (Lunze, 2004b)), for which the sampling time T_S is applied. The control loop is shown in

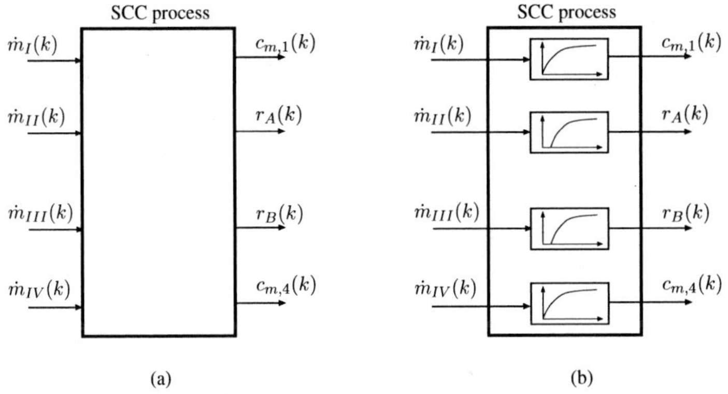

Figure 5.2: Input and output signals (a) and dynamical input–output behaviour (b) of the continuously controlled SCC plant

Figure 5.3. The controllers are indicated by K_i, $i = 1, 2, 3, 4$. The four SISO systems of the plant model are labelled by the time constants T_i and the gain factors k_i, $i = 1, 2, 3, 4$.

The following method is considered for the design of the controllers K_i, $i = 1, 2, 3, 4$: The dynamics of the plant are identified as *continuous–time systems* in form of first–order transfer functions. The time constants and gains of the continuous–time transfer functions

$$G_i(s) = \frac{k_i}{T_i\, s + 1} \cdot \mathrm{e}^{-s\, T_{d,i}} \,,$$

for $i = 1, 2, 3, 4$ are determined from the step response of the considered SCC plant based on numerical simulations and experimental tests. For $i = 1, 4$, the time delay $T_{d,i}$ is zero, and for $i = 2, 3$, the time delay is $T_{d,i} = T_S$. A continuous–time PI controller is designed using the plant model and transformed to a discrete–time control law. For the design of the controller parameters the continuous–time controller transfer function

$$K_i(s) = k_{p,i}\, \frac{T_{I,i}\, s + 1}{T_{I,i}\, s} \tag{5.1}$$

is considered. The integrator time constant $T_{I,i}$ is set to

$$\boxed{T_{I,i} = T_i \,.} \tag{5.2}$$

If the time delay is not considered, the transfer function of each control loop is given by

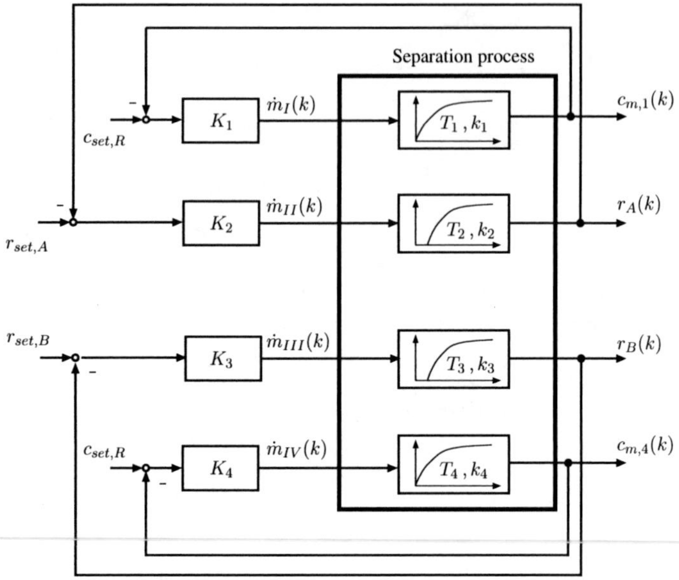

Figure 5.3: Measurement–based continuous SCC control loop

$$G_{w,i} = \frac{Y_i(s)}{W_i(s)} = \frac{1}{T_{w,i}\, s + 1} \,,$$

where Y_i denotes the respective controlled output and W_i denotes the set–point. The time constant of the closed control loop is

$$T_{w,i} = \frac{T_{I,i}}{k_{p,i}\, k_i} \,.$$

Choosing the closed–loop time constant $T_{w,i}$ as the tuning parameter, the following controller gain is obtained:

$$\boxed{k_{p,i} = \frac{T_i}{T_{w,i}\, k_i} \,.} \tag{5.3}$$

The control loop is stable, if $T_{w,i} > T_{d,i}$ is chosen. Hence, for the purity control loops $i = 2, 3$,

$$\boxed{T_{w,i} > T_S} \tag{5.4}$$

must hold.

For the implementation of the controller, the continuous–time transfer function (5.1) is converted to the discrete–time control algorithm

$$\begin{aligned} x_{c,i}(k+1) &= x_{c,i}(k) + k_{p,i}\,\tfrac{T_S}{T_{I,i}}\,e_i(k) \\ u_i(k) &= x_{c,i}(k) + k_{p,i}\left(1 + \tfrac{T_S}{2\,T_{I,i}}\right)e_i(k)\,, \end{aligned} \tag{5.5}$$

where $e_i(k) = w_i(k) - y_i(k)$ is the control offset, $x_{c,i}(k)$ is the controller state and $u_i(k)$ is the controller output.

The concept applies to SMB and VARICOL processes with a time–invariant switching pattern. For the latter case, k has to be replaced by k_S.

5.3 Observation–based control

This section describes a control concept for SCC processes, whose wave front concentrations are determined by wave front observers introduced in Chapter 4. Because the application of the observer was restricted to SMB processes, the hereafter presented concept is only applied to the SMB.

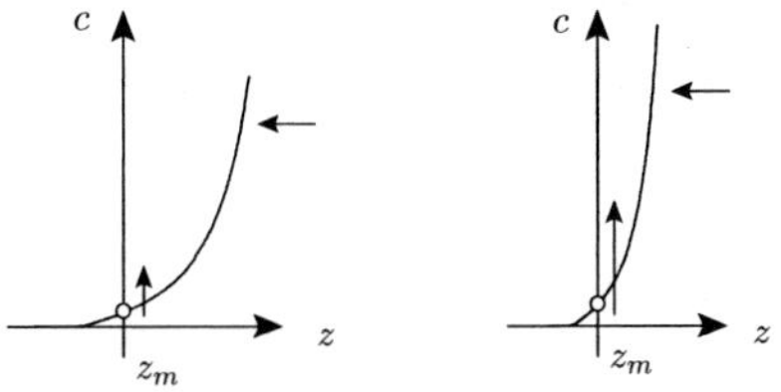

Figure 5.4: Flat and steep wave fronts

5.3.1 Control concept

If the adsorption shows significant nonlinear behaviour or if a low diffusion has to be considered, the concentration profile wave fronts become very steep. Hence, using the feedback of an equidistantly sampled wave front concentration $c_i(\tau_{mi}, z_{mi}, k)$ for the control as considered in Section

5.2 can lead to heavy oscillations. Considering a steep shape of the wave front (Figure 5.4), a disturbance or minimal control input variation leads to a large change of the sampled wave front concentration, which is difficult to handle by a linear controller alone. To overcome this problem it is necessary to include information about the shape and propagation of the wave front into the control loop. This is possible by applying the wave front observation presented in Chapter 4.

The following steps show, how the wave front observer states can be used for this purpose. The first step shows, how the information about the shape and propagation is obtained from the wave front observer. The second step analyses the wave front states for the stationary state and proposes a candidate state as the controlled variable. The third step investigates how this state is influenced by the shape and the propagation in the transient of the wave front and considers the dynamics with respect to the control input.

Step 1: Wave front shape and propagation. For the wave front c_i, the wave front observer provides the states

$$\begin{aligned} x_{i,1}(k) &= b_i(k) \\ x_{i,2}(k) &= -b_i(k)\, v_{c,i}(k) \\ x_{i,3}(k) &= -b_i(k)\, z_{0,i}(k)\,, \end{aligned} \tag{5.6}$$

where b_i describes the shape, $v_{c,i}$ the wave front propagation velocity and $z_{0,i}$ the spatial offset at $\tau = 0$. If a variation of the control input or a disturbance occurs, these parameters perform a transition which is tracked by the observer. Hence, the desired information about the shape and the propagation is provided by the wave front observer.

Step 2: Stationary wave front state $x_{i,3}$. According to the wave front model presented in Section 4.2 the wave front c_i for $\tau \in [0, T_S]$, $z \in [0, L]$ is given by

$$c_i(\tau, z, k) = \mathrm{e}^{b_i(k)\,(z - v_{c,i}(k)\,\tau - z_{0,i}(k))}\,. \tag{5.7}$$

In the stationary state, the propagation velocity of each wave front is

$$v_{c,i} = \frac{L}{T_S}\,. \tag{5.8}$$

The largest impurity by the wave fronts c_i, $i = 1, 2, 3, 4$ at the positions z_A, z_B or z_S is given by

$$c_{max,i}(k) = c_i(T_S, L, k)\,. \tag{5.9}$$

For the stationary state, these wave front concentrations are determined from Equations (5.6), (5.7), (5.8) and (5.9) by

$$\begin{aligned} c_{max,i}(k) &= \mathrm{e}^{b_i(k)\,(L-\frac{L}{T_S}\,T_S-z_{0,i}(k))} \\ &= \mathrm{e}^{-b_i(k)\,z_{0,i}(k)} \\ &= \mathrm{e}^{x_{i,3}(k)}\,. \end{aligned} \tag{5.10}$$

Equation (5.10) is transformed to

$$\ln(c_{max,i}(k)) = x_{i,3}(k)\,,$$

which shows that, in the stationary state, $x_{i,3}(k)$ can be expressed as a function of $c_{max,i}(k)$. In other words, if a controller drives $x_{i,3}(k)$ to the set–point value

$$x_{i,3}(k) = x_{set,i,3} = \ln(c_{set,i})\,, \tag{5.11}$$

then the considered impurity concentration $c_{max,i}$ of the wave front model corresponds to the set–point concentration $c_{set,i}$.

Step 3: Dynamics of $x_{i,3}$. The state–space model (4.11), (4.12) of the wave front states shows that the state $x_{i,3}(k+1)$ is expressed as a linear combination of all actual states $x_{i,1}(k)$, $x_{i,2}(k)$ and $x_{i,3}(k)$ and, hence, includes the information about the shape and the propagation of the wave front:

$$x_{i,3}(k+1) = L\,x_{i,1}(k) + T_S\,x_{i,2}(k) + x_{i,3}(k)\,.$$

This equation shows that, if a transition of the wave front occurs, which induces a change of the shape $b_i(k)$ and the propagation velocity $v_{c,i}(k)$, then the state $x_{i,3}(k)$ will directly be affected and performs a transition. If the controller is able to suppress the transition and bring $x_{i,3}$ back to the set–point according to Equation (5.11), the disturbance is rejected. To be able to apply this concept, the wave front states $x_{i,3}(k)$, $i = 1, 2, 3, 4$ have to be determined by applying wave front observers. Because the dynamical behaviour of $x_{i,3}$ with respect to the control input can be approximated by first–order systems, and because $x_{i,3}$ contains the information about the change of the wave front, it is used as the controlled variable instead of the equidistantly sampled wave front concentration $c_{m,i}$ to control the respective wave front.

5.3.2 Plant model

The concept for the control of the wave fronts discussed in the previous section requires the application of each one wave front observer for the determination of the wave front states $x_{i,3}(k)$. According to the results of Section 4.3, the input signals of the wave front observers are the wave front concentration measurements $c_{m,i}(k) = c_i(\tau_{m,i}(k), z_{m,i}, k)$. Here, $\hat{x}_{i,3}(k)$ is considered as the controlled output signal. The purity values are measured directly on the plant. In terms of the controller design, the plant structure shown in Figure 5.5 (a) is obtained. The analysis of the input–output behaviour with respect to the controlled output signals shows that the cross coupling between the control input $\dot{m}_j(k)$, $j = I, II, III, IV$ and the wave front states $\hat{x}_{i,3}$, $i = 1, 2, 3, 4$ is negligible and each wave front state shows a discrete–time linear first order behaviour with respect to the flow rate of the respective internal fluid. For the purity values, the same dynamical behaviour as described in Section 5.2 is considered. Hence, for the controller design model, the structure shown in Figure 5.5 (b) is obtained.

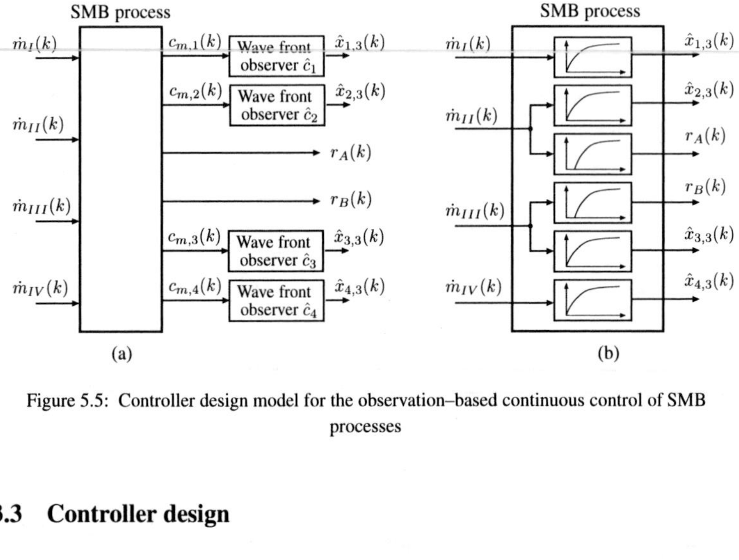

Figure 5.5: Controller design model for the observation–based continuous control of SMB processes

5.3.3 Controller design

The plant which is considered for the observation–based control has four control input and six controlled output signals (Figure 5.5 (b)). The control concept considers the control of the wave front states to a given set–point. The control of the purity values is realised by a cascaded controller of which the outer loop provides the set–points for the wave front states, which are controlled by the inner control loop. The controller structure is shown in Figure 5.6. It consists of

◇ standard control loops for the wave front states $\hat{x}_{1,3}(k)$ and $\hat{x}_{4,3}(k)$, which are controlled by the controllers K_1 and K_4 and

◇ cascaded control loops for the purity values $r_A(k)$ and $r_B(k)$, of which the inner loop controls the wave front states $\hat{x}_{2,3}(k)$ and $\hat{x}_{3,3}(k)$ by the controllers K_2 and K_3, and the outer loop controls the purity values $r_A(k)$ and $r_B(k)$ by the controllers K_A and K_B.

The controller design for the standard control loops is based on the transformation of the problem to the design of a continuous–time controller and follows the procedure proposed in Section 5.2.2. The design of the cascaded controller is also based on the transformation of the plant and the controller model into continuous–time transfer functions. The plant models

$$G_i(s) = \frac{\hat{x}_{i,3}(s)}{\dot{m}_j(s)} = \frac{k_i}{T_i\, s+1} \tag{5.12}$$

for the pairs $(i,j) \in \{(2,II),(3,III)\}$ and

$$G_* = \frac{r_*(s)}{\dot{m}_j(s)} = \frac{k_*}{T_*\, s+1} \cdot \mathrm{e}^{-s\,T_S} \tag{5.13}$$

for the pairs $(*,j) \in \{(A,II),(B,III)\}$ are obtained, where T_S is the switching time. For these two systems the PI controllers

$$K_i(s) = k_{p,i}\,\frac{T_{I,i}\, s+1}{T_{I,i}\, s}\,, \tag{5.14}$$

$$K_*(s) = k_{p,*}\,\frac{T_{I,*}\, s+1}{T_{I,*}\, s}\,. \tag{5.15}$$

for $i = 2,3$ and $* = A,B$ have to be designed.

For the design of the inner loop, the integrator time constant is set to

$$\boxed{T_{I,i} = T_i} \tag{5.16}$$

and the controller gain is set to

$$\boxed{k_{p,i} = \frac{1}{k_i}\,.} \tag{5.17}$$

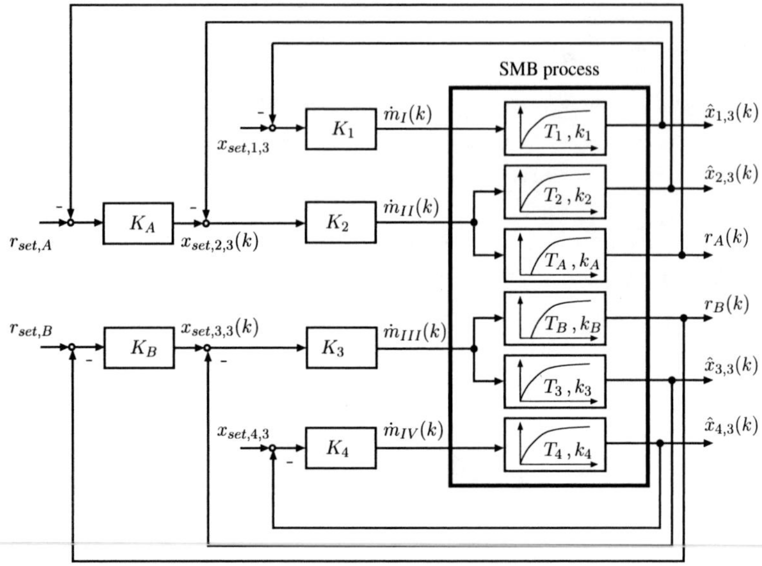

Figure 5.6: Observation–based SMB control loop

Then, the transfer function

$$G_{in,i}(s) = \frac{\dot{m}_j(s)}{x_{set,i,3}(s)} = \frac{K_i(s)}{1 + K_i(s)\, G_i(s)}\,, \tag{5.18}$$

for $(i, j) \in \{(2, II), (3, III)\}$, which describes the dynamics of the inner loop, reduces to

$$G_{in,i} = k_{p,i} = \frac{1}{k_i}\,.$$

These transformation steps are shown in Figure 5.7. Based on this result, the plant model for the outer loop is obtained by applying Equations (5.13) and (5.18):

$$G(s) = \frac{r_*(s)}{x_{set,i,3}(s)} = \frac{1}{k_i}\, G_*(s) = \frac{1}{k_i}\, \frac{k_*}{T_*\, s + 1}\,. \tag{5.19}$$

Based on Equation (5.19), the controller (5.15) of the outer loop is designed using the approach described in Section 5.2.2. For the integrator time constants,

$$T_{I,*} = T_* \tag{5.20}$$

is obtained and for the controller gain

$$k_{p,*} = \frac{k_i}{k_*}\frac{T_*}{T_{w,*}}\,. \tag{5.21}$$

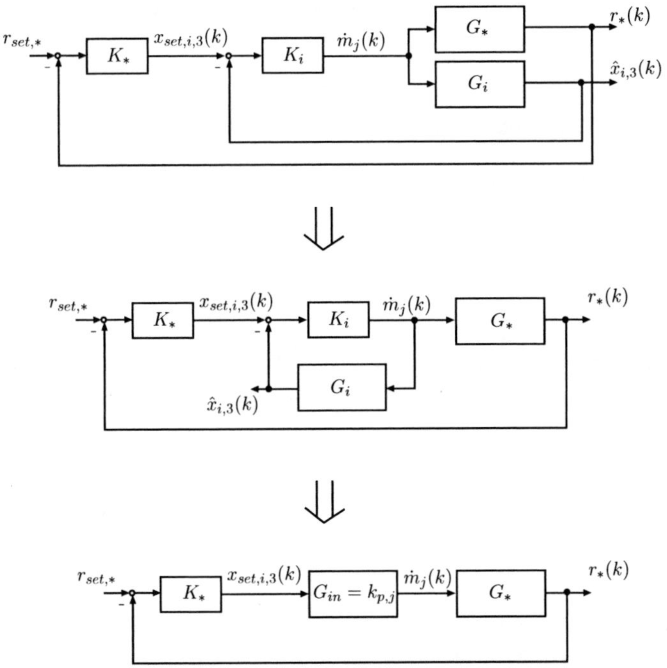

Figure 5.7: Transformation steps for the design of the cascaded purity controller

$T_{w,*}$ is the time constant of the outer control loop, which is the tuning parameter for the controllers K_* with $* = A, B$. To guarantee closed–loop stability, $T_{w,*}$ has to be chosen as

$$T_{w,*} > T_S\,. \tag{5.22}$$

Because the parameters of the controllers K_i, $i = 2, 3$ are determined by the parameters of the models G_i, $i = 2, 3$, and because the time constants of the controllers K_* are determined by the dynamics of the inner control loop, $T_{w,*}$ is the only tuning parameter of the cascaded controller, which renders the tuning very simple.

5.4 Model identification

Modelling method. The input–output behaviour of the considered SCC processes is determined from numerical simulations and experimental tests. Based on the results, the coupling analysis of the stationary and dynamical behaviour is performed and the structure and the parameters of the dynamical model are identified.

The modelling is performed using the step response $h(t)$ and the impulse response $g(t) = \frac{d}{dt}h(t)$. The step response of a MIMO systems is represented by the step response matrix

$$\boldsymbol{H}(t) = \begin{pmatrix} h_{11}(t) & h_{12}(t) & \dots & h_{1n_u}(t) \\ h_{21}(t) & h_{22}(t) & \dots & h_{2n_u}(t) \\ \vdots & \vdots & & \vdots \\ h_{n_y1}(t) & h_{n_y2}(t) & \dots & h_{n_yn_u}(t) \end{pmatrix},$$

where $h_{ij}(t)$ is the step response of the ith measured output y_i, $i = 1, 2, \dots, n_y$ for the jth control input u_j, $j = 1, 2, \dots, n_u$. Correspondingly, the impulse response matrix is

$$\boldsymbol{G}(t) = \begin{pmatrix} g_{11}(t) & g_{12}(t) & \dots & g_{1n_u}(t) \\ g_{21}(t) & g_{22}(t) & \dots & g_{2n_u}(t) \\ \vdots & \vdots & & \vdots \\ g_{n_y1}(t) & g_{n_y2}(t) & \dots & g_{n_yn_u}(t) \end{pmatrix}.$$

Based on the step response matrix $\boldsymbol{H}(t)$, the static coupling is investigated and the dynamical transition is modelled. Based on the impulse response matrix $\boldsymbol{G}(t)$, the coupling which affects the transient behaviour, is investigated. These are standard control engineering procedures (see e.g. (Lunze, 2004a,b)).

Analysis of the static coupling. The step response $h_{ij}(t)$ of a MIMO system is the transient of the ith measured output y_i due to a normalised step of the jth control input u_j. For stable systems, $h_{ij}(t)$ converges to a constant stationary value k_{sij}. The static coupling of the MIMO system is described by the normalised static matrix

$$\bar{\boldsymbol{K}}_s = \frac{1}{k^2_{sij,max}} \begin{pmatrix} k^2_{s11} & k^2_{s12} & \dots & k^2_{s1n_u} \\ k^2_{s21} & k^2_{s22} & \dots & k^2_{s2n_u} \\ \vdots & \vdots & & \vdots \\ k^2_{sn_y1} & k^2_{sn_y2} & \dots & k^2_{sn_yn_u} \end{pmatrix}.$$

Analysis of the dynamical coupling. The elements of the dynamical coupling matrix $\bar{\boldsymbol{K}}_d$ are determined from the impulse response $g_{ij}(t) = \frac{d}{dt} h_{ij}(t)$, which describes the transient of the ith measured output y_i due to an impulse of the jth control input u_j:

$$k_{dij} = \frac{1}{k^2_{sij}} \int\limits_0^\infty g^2_{ij}(t)\, dt\,.$$

The dynamical coupling matrix $\bar{\boldsymbol{K}}_d$ is determined by

$$\bar{\boldsymbol{K}}_d = \frac{1}{k^2_{dij,max}} \begin{pmatrix} k^2_{d11} & k^2_{d12} & \dots & k^2_{d1n_u} \\ k^2_{d21} & k^2_{d22} & \dots & k^2_{d2n_u} \\ \vdots & \vdots & & \vdots \\ k^2_{dn_y1} & k^2_{dn_y2} & \dots & k^2_{dn_yn_u} \end{pmatrix}.$$

System analysis and modelling. The normalised coupling matrices $\bar{\boldsymbol{K}}_s$ and $\bar{\boldsymbol{K}}_d$ allow to quickly evaluate the cross coupling of MIMO systems. If both matrices are diagonally dominant, the cross coupling is negligible. An n–dimensional matrix

$$\boldsymbol{S} = \begin{pmatrix} s_{11} & s_{12} & \dots & s_{1n} \\ s_{21} & s_{22} & \dots & s_{2n} \\ \vdots & \vdots & & \vdots \\ s_{n1} & s_{n2} & \dots & s_{nn} \end{pmatrix}$$

is called diagonally dominant,

$$\begin{aligned} \text{if} \quad & |s_{ii}| > \sum_{i=1, i\neq j}^{n} |s_{ij}| \\ \text{or} \quad & |s_{ii}| > \sum_{j=1, i\neq j}^{n} |s_{ij}| \end{aligned} \tag{5.23}$$

holds (Lunze, 2004b). If one of the matrices $\bar{\boldsymbol{K}}_s$ and $\bar{\boldsymbol{K}}_d$ is diagonally dominant, the considered system can be controlled by decentralised controllers. For the cross coupling analysis of SCC processes, this approach is applied.

For all SCC processes and separation examples considered in this chapter, a low static cross coupling occurs, which will be shown for each example in Section 5.5. Furthermore, the step responses show the behaviour of first order systems. Hence, the modelling approach is to apply linear discrete–time first order models and identify the model parameters.

5.5 Applications

5.5.1 Separation examples and plant configuration

The control concepts detailed in Section 5.2 and 5.3 are applied to three example separations which show different properties with respect to the adsorption behaviour and the plant configuration. With respect to the adsorption behaviour,

- ◇ the Langmuir–like adsorption (S- and R-Ibuprofen), and
- ◇ the anti–Langmuir–like adsorption (α– and δ–Tocopherol)

are considered. With respect to the plant configuration, the following cases are regarded:

- ◇ a six column SMB process with the configuration (1/2/2/1),
- ◇ an eight column SMB process with the configuration (2/2/2/2) and
- ◇ a four column VARICOL process with the configuration $(0{,}58/1{,}17/1{,}54/0{,}71)$.

	Mixture	Adsorption	Plant	n_c	Configuration
1.	S-, R-Ibuprofen	Langmuir	SMB	6	$(1/2/2/1)$
2.	α–, δ–Tocopherol	Anti–Langmuir	SMB	8	$(2/2/2/2)$
3.	S-, R-Ibuprofen	Langmuir	VARICOL	4	$(0{,}58/1{,}17/1{,}54/0{,}71)$

Table 5.1: Example processes for continuous SCC control

Table 5.1 summarises the considered combinations of separation mixtures and plant configurations. For each example, the following design steps are presented:

1. description of the control problem for the considered example,
2. separate numerical simulation (and, if available, experimental test) of the step response for each control input,

3. analysis of the static and dynamic cross coupling,

4. identification of the controller design model,

5. controller design,

6. investigation of the closed–loop behaviour with respect Task 5.1.1.

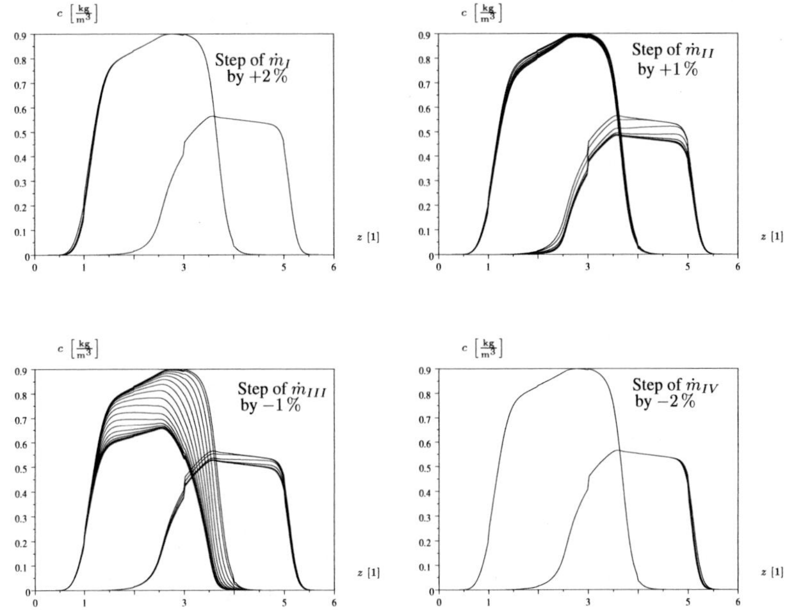

Figure 5.8: Snapshots of the Ibuprofen concentration profile step response

5.5.2 Ibuprofen separation in a 6–column SMB plant

Plant specifications. The mixture of R– and S–Ibuprofen in supercritical CO_2 and Isopropanol shows a Langmuir–like adsorption behaviour. The plant parameters are given in Appendix B. The operation point parameters are listed in Table B.3.

Controller scheme. For the control of this example process it is assumed that the concentrations of the wave fronts c_1 and c_4 are measured directly at the position z_S and the measurement time is

$\tau_m = \frac{1}{2} T_S$. The measurement–based control concept presented in Section 5.2 is applied (Figure 5.3).

Modelling. For the analysis of the cross coupling and the identification of the dynamics, numerical simulations of the step responses are used. Figure 5.8 shows snapshots of the concentration profile step response. The following port positions are considered: $z_S = 0$, $z_A = 1$, $z_{A+B} = 3$, $z_B = 5$, $z_R = 6$. Each of the plots shows the response of the internal concentration profiles for the step of one internal fluid flow rate. The snapshots are recorded after each cycle K at the local time $\tau = \frac{1}{2} T_S$. From the plots it can be concluded that each wave front is mainly influenced by the corresponding internal fluid flow rate of the section, through which it propagates and, hence, the cross coupling is expected to be low.

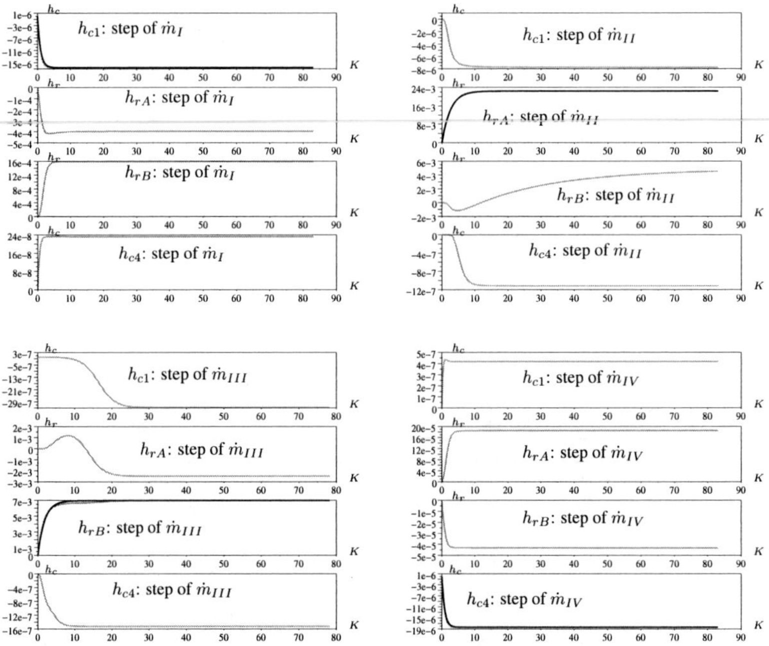

Figure 5.9: Step response and model of $c_{m,1}(k)$, $r_A(k)$, $r_B(k)$ and $c_{m,4}(k)$ of the R– and S–Ibuprofen separation in a 6–column SMB plant

Analysis of the cross coupling and model identification. Figure 5.9 shows the response of the output variables $c_{m,1}$, r_A, r_B and $c_{m,4}$ for normalised steps of the internal fluid flow rates by an amount of $d\dot{m}_j = 0{,}01 \cdot \dot{m}_{0,j}$, $j = I, II$, i.e. a step of $+1\%$, and $d\dot{m}_j = -0{,}01 \cdot \dot{m}_{0,j}$, $j = III, IV$, i.e. a step of -1% (gray line).

The results of the cross coupling analysis are given by the static coupling matrix $\bar{\boldsymbol{K}}_s$ and the dynamical coupling matrix $\bar{\boldsymbol{K}}_d$ in the following equations.

$$\bar{\boldsymbol{K}}_s = \begin{pmatrix} 0.0055 & 0.0012 & 0.0002 & 0 \\ 0.0003 & 1 & 0.0122 & 0.0001 \\ 0.0051 & 0.041 & 0.0971 & 0.0001 \\ 0 & 0 & 0 & 0.0004 \end{pmatrix}$$

$$\bar{\boldsymbol{K}}_d = \begin{pmatrix} 0.26 & 0.12 & 0.04 & 1 \\ 0.28 & 0.1 & 0.13 & 0.17 \\ 0.16 & 0.03 & 0.15 & 0.27 \\ 0.45 & 0.09 & 0.09 & 0.29 \end{pmatrix}$$

Each of the columns of $\bar{\boldsymbol{K}}_s$ refers to one of the four plots in Figure 5.9. The static coupling matrix $\bar{\boldsymbol{K}}_s$ is diagonally dominant according to Equation (5.23). The dynamical coupling matrix $\bar{\boldsymbol{K}}_d$ is not diagonally dominant. This means that each input causes a transient of all outputs, which, however, can be neglected because of the low static cross coupling.

The model parameters determined from the identification of the step responses $h_{ii}(t)$, $i = 1, 2, 3, 4$ represented in Figure 5.9 (black lines) are listed in Table 5.2. The gains are determined for the following units: c_1, c_4 in $\left[\frac{\text{kg}}{\text{m}^3}\right]$, and r_A, r_B in $[\%]$. The controller output is considered as

$$d\dot{m}_j(k) = \dot{m}_j(k) - m_{0,j}\,, \quad j = I, II, III, IV\,,$$

with the unit $\left[\frac{\text{kg}}{\text{s}}\right]$.

Gains	k_1	k_2	k_3	k_4
	$-7{,}86$	$13{,}5$	-4	13
Time constants	T_1	T_2	T_3	T_4
in [s]	664	1805	1359	596
in cycles	0,922	2,51	1,89	0,83

Table 5.2: Model parameters for the SMB Ibuprofen separation

Controller design and closed–loop behaviour. Because low cross couplings are considered, the decentralised discrete–time PI controller scheme proposed in Section 5.2 is applied. The time constants of the controllers K_i, $i = 1, 2, 3, 4$ are chosen according to Equation (5.2). For the closed–loop time constant, $T_{w,i} = 14400$ s, which corresponds to 20 cycles, is chosen. The resulting controller gains are determined using Equation (5.3) and are given in Table 5.3.

Gains	$k_{p,1}$	$k_{p,2}$	$k_{p,3}$	$k_{p,4}$
	$-5{,}9 \cdot 10^{-3}$	$9{,}3 \cdot 10^{-3}$	$-23{,}6 \cdot 10^{-3}$	$3{,}2 \cdot 10^{-3}$

Table 5.3: Controller gains for the SMB Ibuprofen separation

The set–point for the wave front concentrations and the purity values are chosen to

$$\begin{aligned} c_{set,1} &= 1 \cdot 10^{-3}\ \frac{\mathrm{kg}}{\mathrm{m}^3}\,, \\ c_{set,4} &= 1 \cdot 10^{-3}\ \frac{\mathrm{kg}}{\mathrm{m}^3}\,, \\ r_{set,A} &= 99{,}9\,\%\,, \\ r_{set,B} &= 99{,}9\,\%\,. \end{aligned}$$

In the following, numerical simulations of the closed–loop dynamical behaviour are presented.

Figures 5.10 and 5.11 show the start–up behaviour of the 6–column SMB plant. Figure 5.10 shows snapshots of the concentration profiles c_A and c_B during the start–up of the plant. The snapshots are recorded every cycle K at $\tau = \frac{1}{2}\,T_S$. The plot shows how the concentration profiles evolve until the stationary state is reached. It also shows how the wave fronts propagate outwards in the direction of the product outlets and the recycling port.

The left plots in Figure 5.11 show the evolution of the control input of the plant during the start–up. The signals are plotted over the numbers of cycles K. The plots show that the internal fluid flow rates are reduced in the sections I and II, and increased in the sections III and IV.

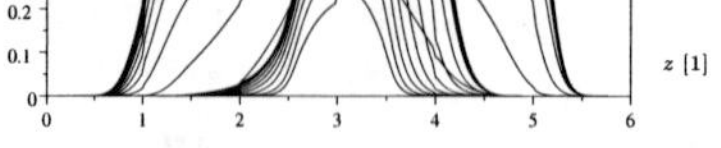

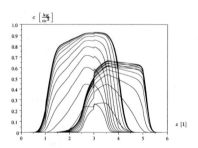

Figure 5.10: Concentration profiles during the start–up of the 6–column SMB plant

In the right plots in Figure 5.11 the progression of the controlled variable is represented. The upper plot shows the measured wave front concentrations $c_{m,1}$ and $c_{m,4}$. The lower plot shows

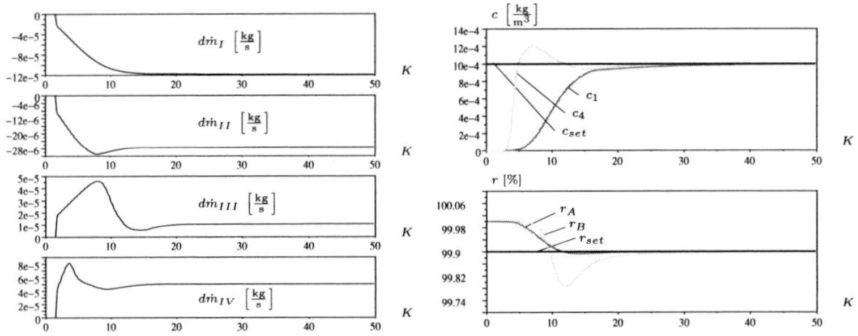

Figure 5.11: Start–up behaviour of the 6–column SMB plant

the purity values r_A and r_B. From both plots it can be seen that the controller drives the plant to the desired set–point.

Figures 5.12 and 5.13 show the dynamical transition of the 6–column SMB plant during the transient caused by the step disturbance of the feed inlet concentration by $+100\%$. Figure 5.12 again shows snapshots of the concentration profiles for every cycle of the transition. The plot shows how the profiles grow to larger concentration values due to the increased mass throughput. It also shows that the position of the wave fronts change marginally.

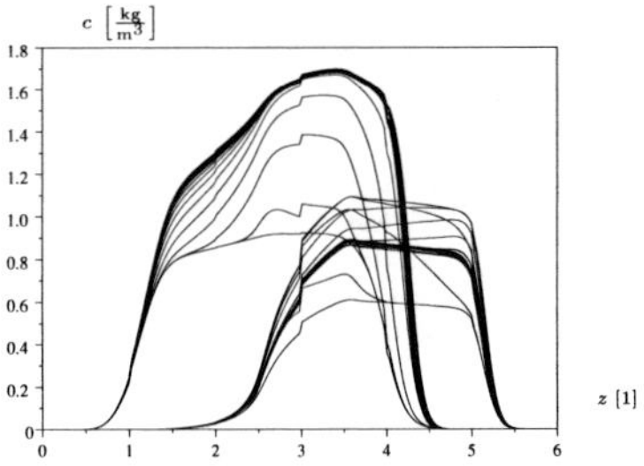

Figure 5.12: Concentration profiles during the rejection of a step disturbance of the feed inlet concentration by 100 % of the 6–column SMB plant

The left plots in Figure 5.13 show how the controller output reacts to the disturbance. The fluid flow in section I is increased to prevent the wave front c_1 from propagating into section IV. In section IV the fluid flow is reduced in total to prevent the wave front c_4 from propagating into section I. In both middle section the internal fluid flow is reduced. This leads to a shift of the wave front c_2 towards z_A.

Because in the area of z_A the product concentration increases, the by–product concentration of

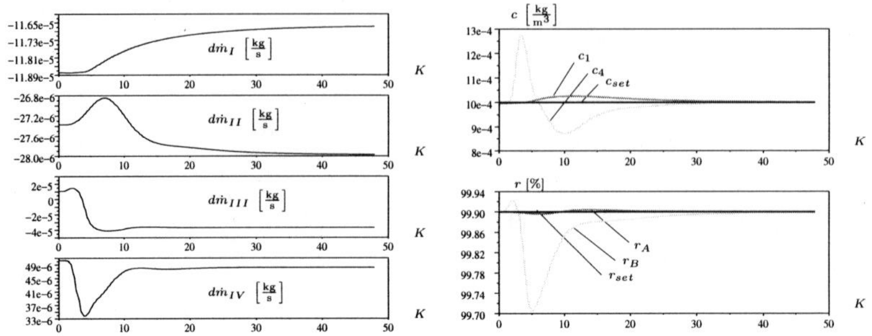

Figure 5.13: Rejection of a step disturbance of the feed inlet concentration by 100 % of the 6–column SMB plant

c_2 is also increased to achieve the required purity value. The set–point values are reached after about 30 cycles (right plots in Figure 5.13).

For a further disturbance analysis, the drift of the column package porosity ε while keeping a constant package volume is considered. It is assumed that the disturbance occurs for all separation columns of the plant in the same manner. Figure 5.14 shows the progression of ε during the simulation. During the first 10 cycles the porosity reduces by 0,75 %. Then, it is kept constant.

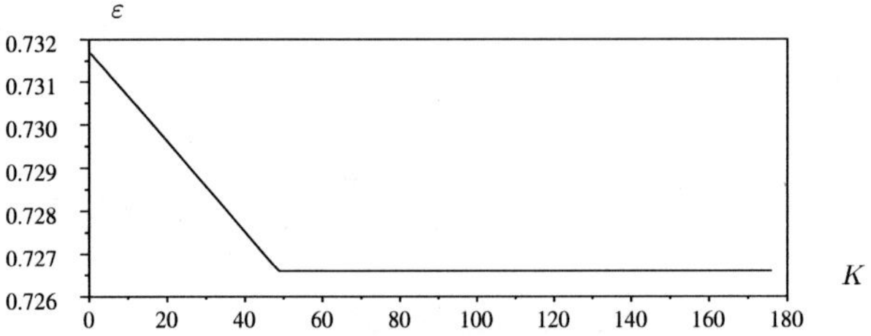

Figure 5.14: Ramp disturbance of the package porosity

Due to the reduced package porosity the internal fluid flow velocity increases and the length of the package inside of the columns decreases. Hence, the propagation of the complete concentration profiles c_A and c_B is accelerated in the direction of the internal fluid flow. To keep the set–points of the recycling impurity and the product purity values, all internal solvent mass flow rates have to be reduced.

Figures 5.15 and 5.16 show the transient behaviour of the 6–column SMB plant during the disturbance. Figure 5.15 shows the snapshots of the concentration profiles. It shows that the shape and position remains almost unchanged. The left plots in Figure 5.16 show the controller input to the plant. During the ramp disturbance, the controller decreases the internal fluid flow rates. The

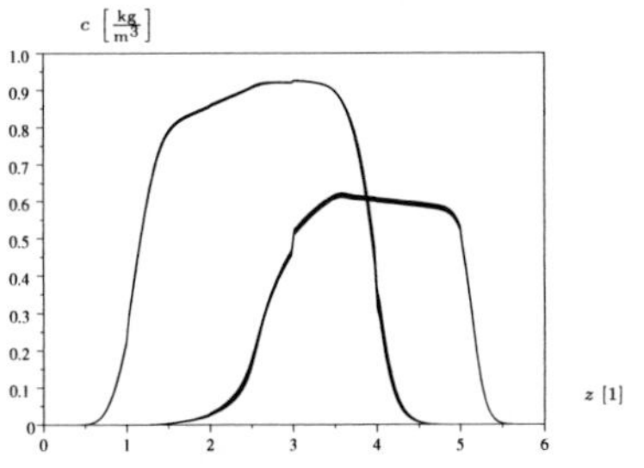

Figure 5.15: Concentration profiles during the transition of the 6–column SMB plant due to the ramp disturbance of the package porosity ε

right plots shows that during the ramp disturbance, a stationary controller error occurs. When the ramp disturbance stops, the recycling impurity and the product purity values are met again after a transient of about 30 cycles (right plots in Figure 5.16).

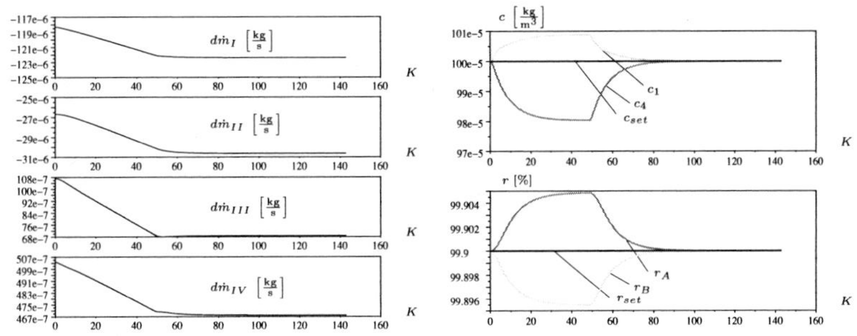

Figure 5.16: Transition of the 6–column SMB plant during the ramp disturbance of the package porosity ε

5.5.3 Tocopherol separation in an 8–column SMB plant

Plant specifications. The mixture of α– and δ–Tocopherol in supercritical CO_2 and Isopropanol shows a Anti–Langmuir–like adsorption behaviour. The plant parameters are given in Appendix B. The operation point parameters are listed in Table B.6.

Controller scheme. Because of the non–linear adsorption and a low diffusion, the observation–based control concept presented in Section 5.3 (Figure 5.6) is applied for the control of this example process. Hence, the modelling of the input–output behaviour of the wave front states $\hat{x}_{i,3}$ and the purity values r_* is necessary.

Modelling. The derivation of the controller design model is performed based on numerical simulations. For the verification of the dynamical behaviour, an experimental test was performed with the real plant. In the following, the numerical simulation results are analysed first. Figure 5.17 shows the step response of the concentration profiles as snapshots, which were recorded as in the previous SCC control example (Section 5.5.2). The port positions are $z_S = 0$, $z_A = 2$, $z_{A+B} = 4$, $z_B = 6$ and $z_R = 8$. From the plots it can be concluded that the cross coupling is very low.

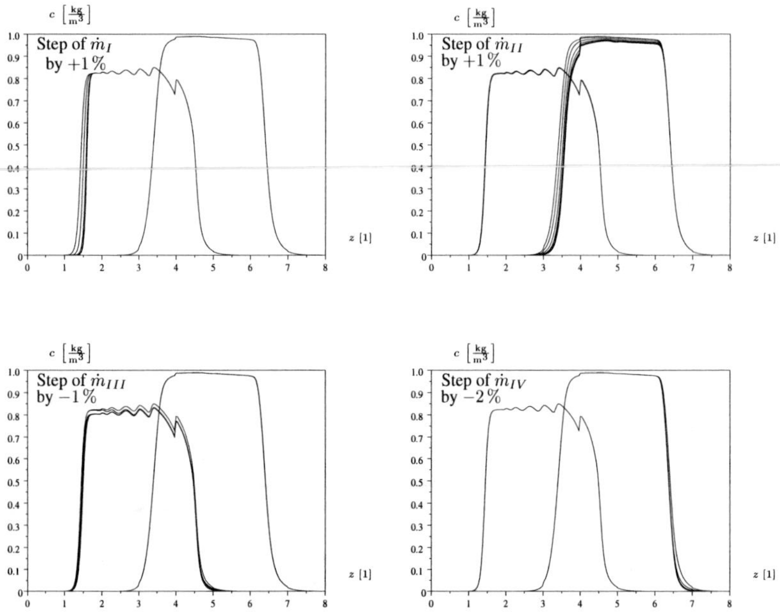

Figure 5.17: Snapshots of the α– and δ–Tocopherol concentration profile step response

Analysis of the cross coupling and model identification. As in the previous SCC control example, the cross coupling is investigated based on the step responses, which are shown in Figure 5.18 for the wave front states $\hat{x}_{i,3}$, $i = 1, 2, 3, 4$. The results of the cross coupling analysis is given by the matrices $\bar{\boldsymbol{K}}_{s,x3}$ and $\bar{\boldsymbol{K}}_{d,x3}$ in the following equations.

$$\bar{\boldsymbol{K}}_{s,x3} = \begin{pmatrix} 1 & 0.0011 & 0.0163 & 0 \\ 0 & 0.0642 & 0 & 0 \\ 0 & 0 & 0.0716 & 0 \\ 0 & 0 & 0 & 0.0323 \end{pmatrix}$$

$$\bar{\boldsymbol{K}}_{d,x3} = \begin{pmatrix} 0.38 & 0.09 & 0.15 & 0.11 \\ 0.61 & 0.31 & 0.12 & 0.48 \\ 0.32 & 0.25 & 0.23 & 0.71 \\ 1 & 0.19 & 0.26 & 0.34 \end{pmatrix}$$

As in the previous SCC control example, the static coupling is low. Hence, the assumption for low cross coupling holds.

For the purity values, the cross coupling between the input signals $\dot{m}_{II}$ and $\dot{m}_{III}$ and the output signals r_A and r_B are investigated. Figure 5.19 shows the step responses for normalised steps of $+1\%$ for $h_{r,A}$ and -1% for $h_{r,B}$ of the respective flow rates. The results of the analysis are represented by the matrices $\bar{\boldsymbol{K}}_{s,r}$ and $\bar{\boldsymbol{K}}_{d,r}$ in the following equations.

$$\bar{\boldsymbol{K}}_{s,r} = \begin{pmatrix} 0.0004 & 0.0001 \\ 0.0004 & 1 \end{pmatrix}$$

$$\bar{\boldsymbol{K}}_{d,r} = \begin{pmatrix} 0.12 & 0.16 \\ 0.06 & 0.1 \end{pmatrix}$$

From the result it is concluded that the cross coupling for the considered input and output signals is low.

Based on the step responses represented in Figures 5.18 and 5.19, the controller design models (Equations (5.12) and (5.13)) are identified. The results are summarised in Table 5.4.

Gains	k_1	k_2	k_3	k_4	k_A	k_B
	$-1{,}01 \cdot 10^6$	$-3{,}4 \cdot 10^5$	$3{,}5 \cdot 10^3$	$3{,}4 \cdot 10^5$	$1{,}9$	$-9{,}2 \cdot 10^2$
Time constants	T_1	T_2	T_3	T_4	T_A	T_B
in [s]	4595	5486	7343	4963	3040	2990
in cycles	2,39	2,85	3,82	2,58	1,58	1,56

Table 5.4: Model parameters for the SMB Tocopherol separation

For the verification of the dynamical behaviour of the wave front concentrations an experimental test was performed with the SFC SMB plant described in Section 2.7. A simultaneous step of all internal fluid flow rates was applied to the plant. Using the discrete concentration profile measurement unit, the wave front concentrations were recorded once per cycle at the positions

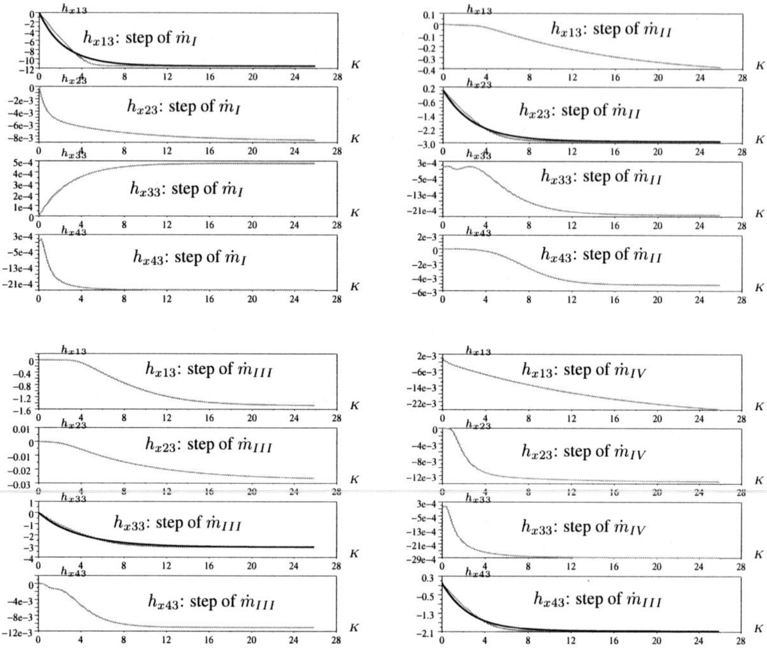

Figure 5.18: Step response and model of $\hat{x}_{i,3}$, $i = 1, 2, 3, 4$ of the α– and δ–Tocopherol separation in an 8–column SMB plant

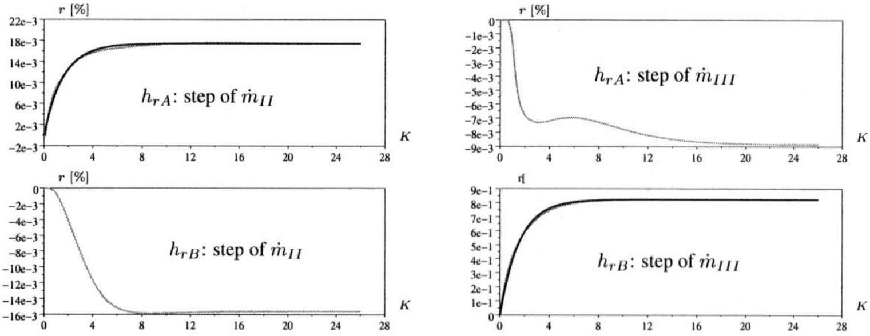

Figure 5.19: Step response and model of r_A and r_B of the α– and δ–Tocopherol separation

$z_{m,1} = 1$, $z_{m,2} = 3$ and $z_{m,3} = 5$ at $\tau_m = \frac{1}{2} T_S$. Because the wave front c_4 had a zero concentration at the position $z = 7$, the position $z_{m,4} = 6$ was chosen for the evaluation of the experiment. The resulting step responses are presented in Figure 5.20. In Appendix C.5 the complete experimental data and the measured concentration profiles for the considered experiment are shown. The signal of the UV detector was not normalised to concentration values. However, it is proportional to the measured concentration. In Figure 5.20, the signal is plotted over K.

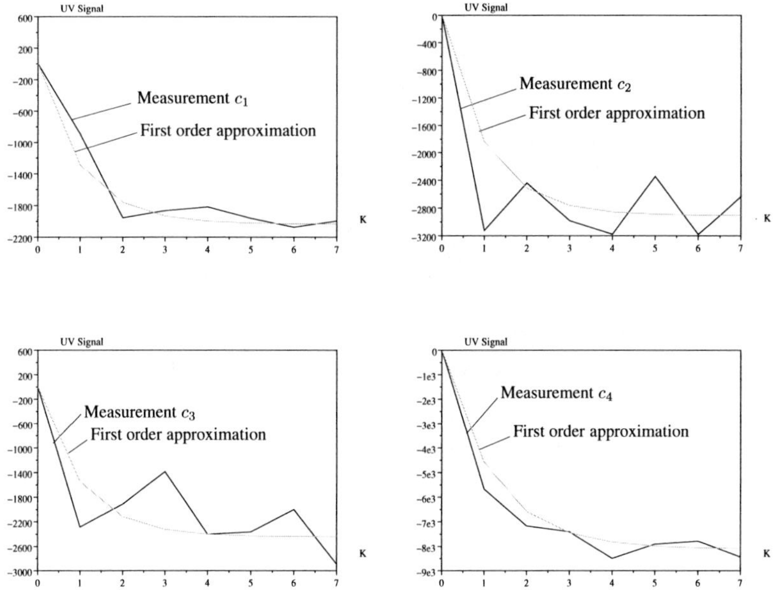

Figure 5.20: Experimental wave front step response

The results in Figure 5.20 show that, on the one hand, a certain variance of the measurement data has to be considered, which is caused by the oscillating pressure profile of the plant and the difficultly to obtain an equilibrated flow of the supercritical fluid. On the other hand, the plots show a clear tendency of the signal evolution which allows for the approximation by first order systems. From these measurements it is concluded that the assumption for the controller design models with respect to the dynamical behaviour is met for practical applications.

Controller design, observer design and closed–loop behaviour. The controller is designed based on the method proposed in Section 5.3.3. For the tuning parameters of the four control

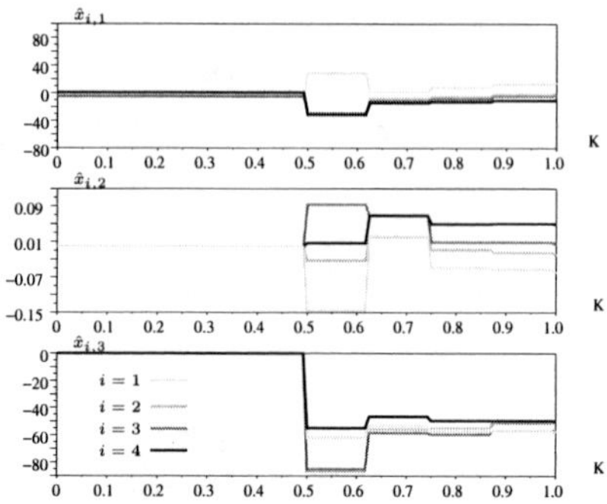

Figure 5.21: Transient of the wave front states after the activation of the wave front observers

loops the same value was chosen, namely $T_{w,i} = T_{w,*} = 3840$ was chosen, which corresponds to a duration of two cycles. Hence, the stability requirement $T_{w,*} > T_S$, $* = A, B$ is met. All resulting controller parameters are listed in Table 5.5.

Gains	$k_{p,1}$	$k_{p,2}$	$k_{p,3}$	$k_{p,4}$	$k_{p,A}$	$k_{p,B}$
	$-1{,}2 \cdot 10^{-6}$	$-2{,}9 \cdot 10^{-6}$	$-2{,}9 \cdot 10^{-6}$	$-3{,}8 \cdot 10^{-6}$	$-14 \cdot 10^{3}$	$0{,}3 \cdot 10^{3}$

Table 5.5: Controller parameters for the Tocopherol separation control

For the set–point, the following values are chosen:

$$
\begin{aligned}
c_{set,1} &= 1 \cdot 10^{-3}\ \frac{\mathrm{kg}}{\mathrm{m}^3} \\
c_{set,4} &= 1 \cdot 10^{-3}\ \frac{\mathrm{kg}}{\mathrm{m}^3} \\
r_{set,A} &= 99{,}98\ \% \\
r_{set,B} &= 99{,}98\ \%\,.
\end{aligned}
$$

For the design of the wave front observers, the measurement times for each wave front have to be chosen. Table 5.6 shows the chosen values. For the error dynamics, the values proposed in Equation (4.32) were chosen.

Wave front	c_1		c_2		c_3		c_4	
Measurement times	τ_{m1}	τ_{m2}	τ_{m1}	τ_{m2}	τ_{m1}	τ_{m2}	τ_{m1}	τ_{m2}
in [s]	180	225	120	225	15	120	15	120

Table 5.6: Measurement times of the wave front observers for the Tocopherol separation control

To validate the closed–loop behaviour the following numerical simulations were performed:

1. The SMB plant performed an open–loop start–up.

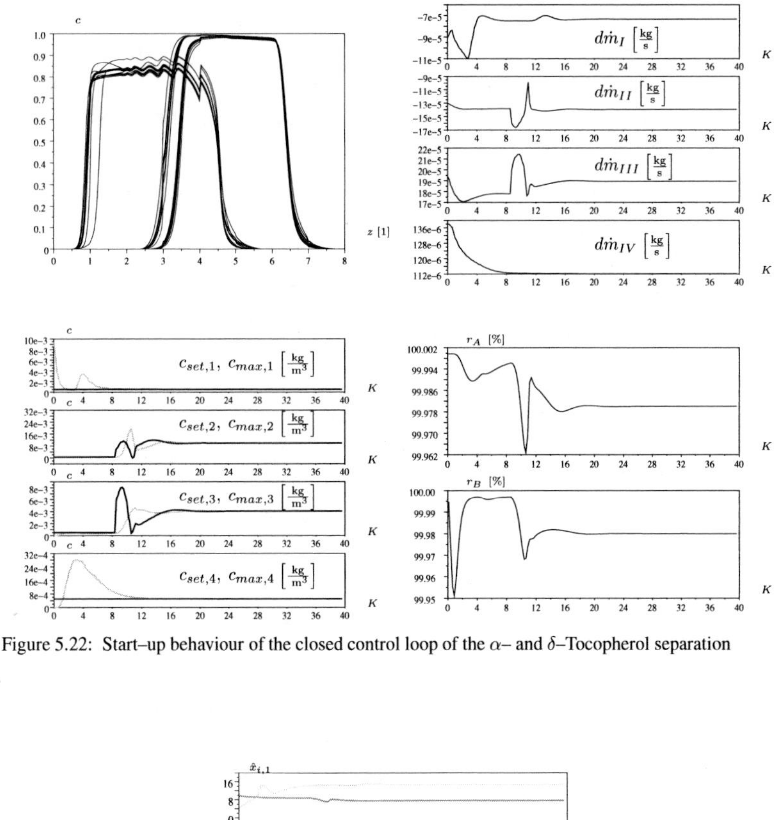

Figure 5.22: Start–up behaviour of the closed control loop of the α– and δ–Tocopherol separation

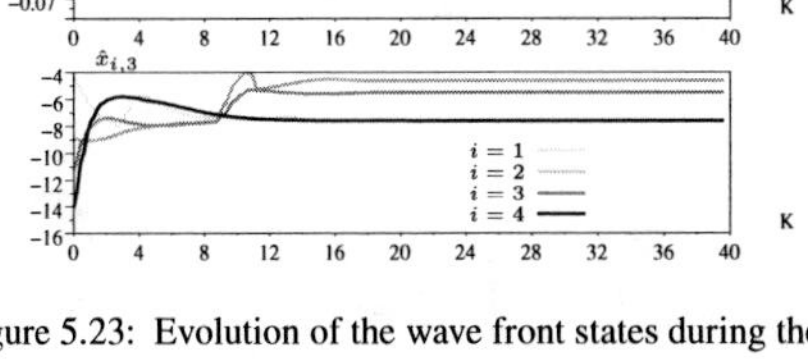

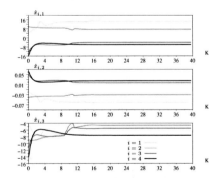

Figure 5.23: Evolution of the wave front states during the transient of the closed control loop

2. After four periods (at $K = 0{,}5$), the wave front observer is activated. Figure 5.21 shows how the wave front states evolve. The inner control loop of the wave front states $\hat{x}_{i,3}(k)$, $i = 1, 2, 3, 4$ is activated after 8 periods at $K = 1$. The outer control loop is then activated after 8 cycles. Figure 5.22 shows four plots of the process variables during the start–up. The upper left plot shows the snapshots of the concentration profiles. The upper right plot shows the evolution of the controller output of the inner control loop, i.e. the adaptation of the internal fluid flow rates.

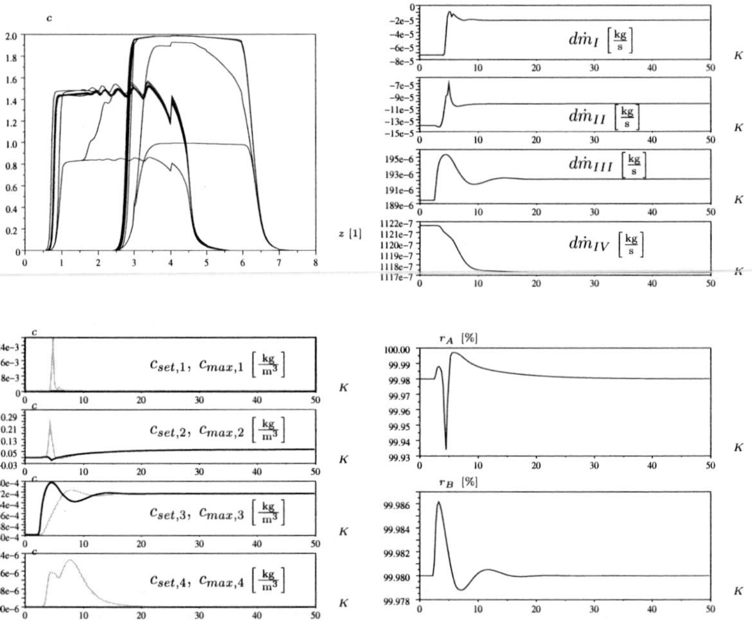

Figure 5.24: Disturbance rejection of a feed input concentration step of the α– and δ–Tocopherol separation

The lower left plot shows the evolution of the set–points (black line) and the recycling and product impurity concentrations (gray lines). While the set–points $c_{set,1}$ and $c_{set,4}$ for $c_{max,1}(z_S, k)$ and $c_{max,4}(z_S, k)$ are constant, the set–points for $c_{max,2}(z_S, k)$ and $c_{max,3}(z_S, k)$ are prescribed by the controllers of the outer control loop. The lower right plot shows the evolution of the purity values $r_A(k)$ and $r_B(k)$. It can be seen from the plots that the corresponding set–point is reached after about 10 cycles after activating the respective controller. Figure 5.23 shows the evolution of the wave front observer states $\hat{x}_{i,1}(k)$, $\hat{x}_{i,2}(k)$ and $\hat{x}_{i,3}(k)$ during the controlled start–up of the plant.

3. In the stationary state, a step disturbance of the feed inlet concentration by 100 % is applied. This is an alternative scenario to point 3. Figure 5.24 shows the evolution of the process variables during the disturbance rejection. The plots show that the set–points of the controlled signals are reached again after about 15 cycles. Figure 5.25 shows the evolution of the wave front states.

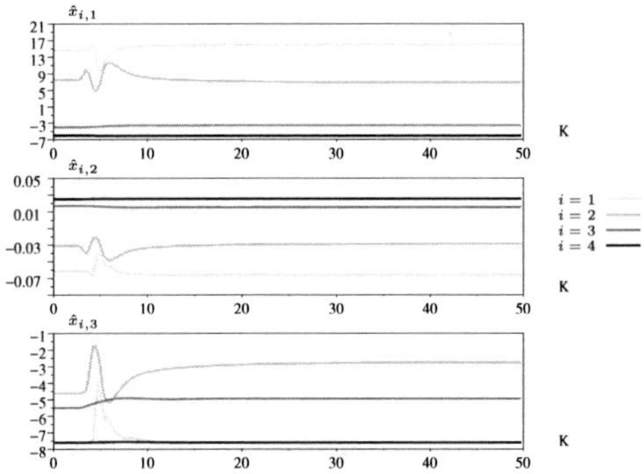

Figure 5.25: Evolution of the wave front states during the rejection of the feed input concentration step

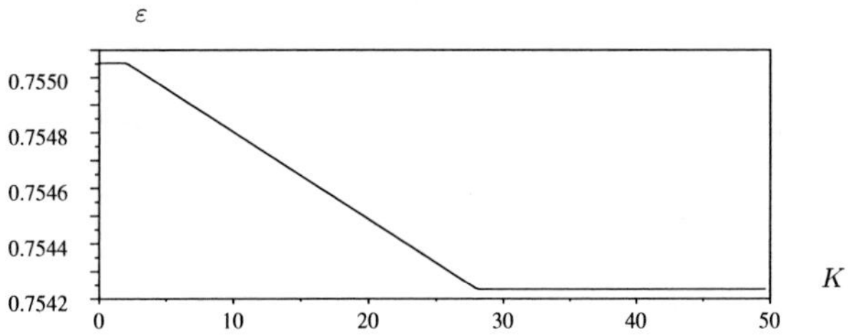

Figure 5.26: Ramp disturbance of the package porosity

4. From the same stationary state which was reached after the controlled start–up of the plant, a ramp disturbance of the package porosity of all column packages is applied. Figure 5.26 shows the evolution of the package porosity during the simulation. Figure 5.27 shows the evolution of the process parameters during the ramp disturbance. Compared to the feed inlet disturbance, the change in the concentration profiles is low and, hence, is the change of the wave front states. Therefore, the wave front states are not presented for this example. The plots in Figure 5.27 show that a control offset occurs for the recycling impurity concentrations, which disappears after the porosity takes a constant value. Because of the cascaded control of the product purity values r_A and r_B, no control offset occurs for these signals.

The plots show, that the set–points are reached after about 10 cycles, which is faster than in the case of the example of the measurement–based control presented in Section 5.5.2.

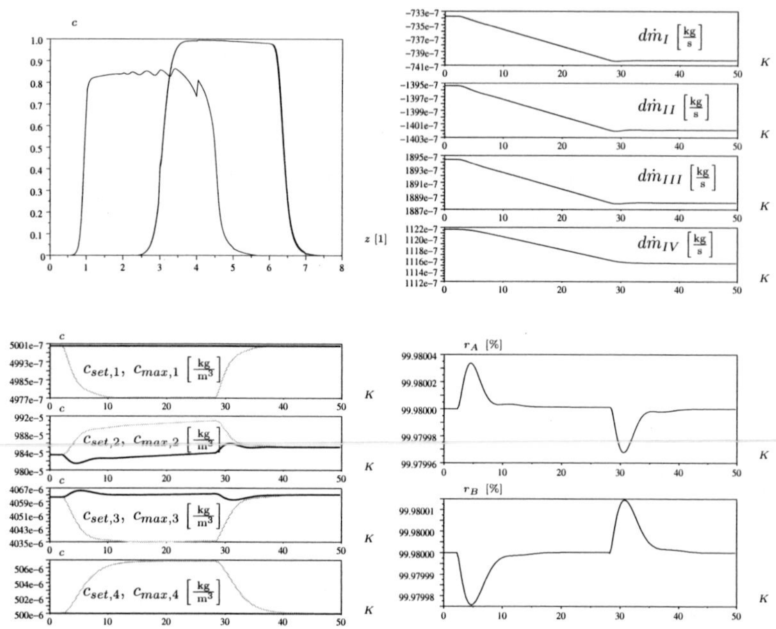

Figure 5.27: Disturbance rejection of a feed input concentration step of the α– and δ–Tocopherol separation

The simulation studies show the potential of the observation–based control approach for practical applications. For the separation example considered in this section, low closed–loop time constants could be applied which led to a considerably fast convergence of the controlled signals to the respective set–points for the considered scenarios. Furthermore, by applying a cascaded controller for the product purity values, no control offset occurs in case of a package porosity drift. However, compared to the measurement–based control concept, the observation–based control approach requires three further discrete–time wave front concentration measurement units.

5.5.4 Ibuprofen separation in a 4–column VARICOL plant

Plant specifications. For this separation example, the same mixture of R– and S–Ibuprofen in the same plant as considered in Section 5.5.2 is regarded. However, it is assumed that the SCC unit has 4 separation columns and is operated as a VARICOL process with a time–invariant switching pattern and the configuration $(0{,}58/1{,}17/1{,}54/0{,}71)$. The internal fluid flow rates of the initial operation point are listed in Table 5.7. The relative switching times are $\Delta T_{A+B} = 30$, $\Delta T_A = 50$ and $\Delta T_B = 85$ and the initial column configuration is $(0/1/2/1)$.

T_S [s]	$\dot{m}_{0,I}$ $\left[\frac{\text{kg}}{\text{s}}\right]$	$\dot{m}_{0,II}$ $\left[\frac{\text{kg}}{\text{s}}\right]$	$\dot{m}_{0,III}$ $\left[\frac{\text{kg}}{\text{s}}\right]$	$\dot{m}_{0,IV}$ $\left[\frac{\text{kg}}{\text{s}}\right]$	$c_{A+B,*}$ $\left[\frac{\text{kg}}{\text{m}^3}\right]$
120	$2{,}33 \cdot 10^{-3}$	$1{,}69 \cdot 10^{-3}$	$1{,}76 \cdot 10^{-3}$	$1{,}17 \cdot 10^{-3}$	1

Table 5.7: Initial operation point for the VARICOL Ibuprofen separation

Controller scheme. With respect to the control it is supposed that the wave front concentrations $c_{m,1}(z_S, k_S)$ and $c_{m,4}(z_S, k_S)$ at the solvent inlet and the purity values $r_A(k_S)$ and $r_B(k_S)$ can be measured directly and are provided at the end of the switching period k_S or $k_S + 1$, respectively. Hence, the plant exhibits the same input–output behaviour as the SMB plant considered in Section 5.5.2.

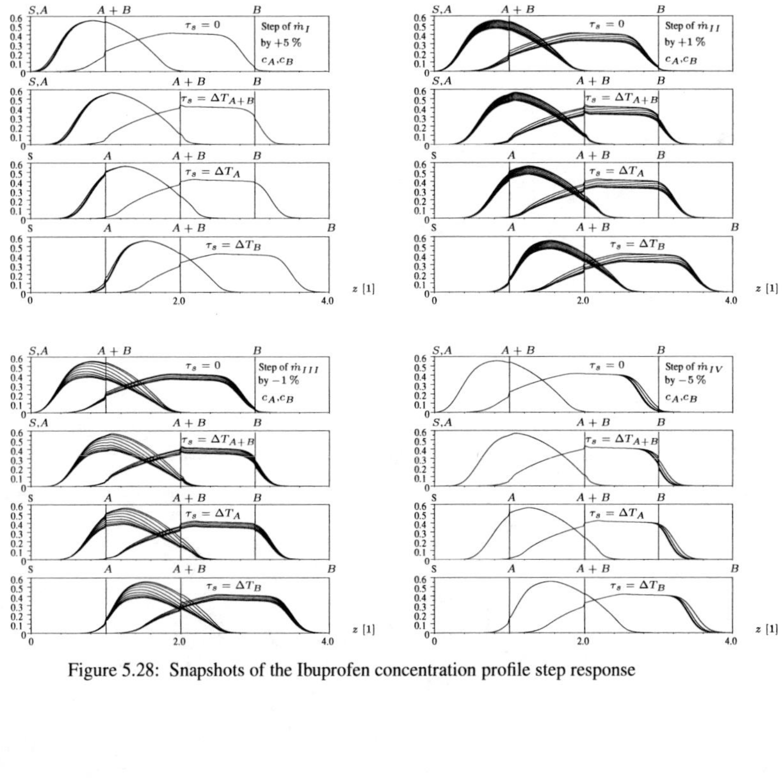

Figure 5.28: Snapshots of the Ibuprofen concentration profile step response

Modelling. Numerical simulations are used to obtain the plant step responses. Figure 5.28 shows the step response of the concentration profiles. It shows that the cross couplings will be low for the the recycling impurity concentrations. However, from the upper right and the lower left plot it can be seen that a certain coupling between the control input signals and the product purity values has to be expected.

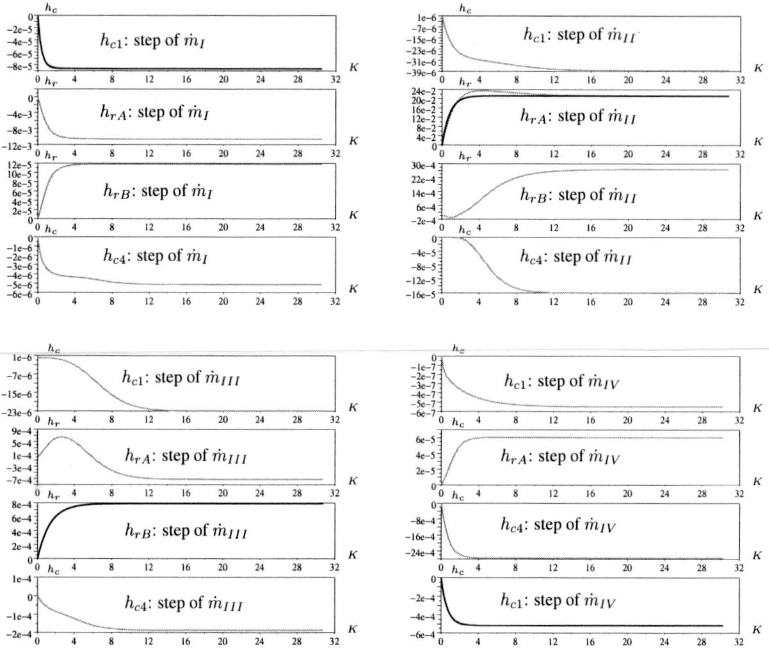

Figure 5.29: Step response and model of $c_{max,i}$, $i = 1,4$ and r_*, $* = A, B$ of the R– and S–Ibuprofen separation in a VARICOL plant

Analysis of the coupling and model identification. The coupling analysis is based on the step responses shown in Figure 5.29. The matrices $\bar{\boldsymbol{K}}_s$ and $\bar{\boldsymbol{K}}_d$ in the following equations indicate the static and dynamical cross coupling.

$$\bar{\boldsymbol{K}}_{s,cmax} = \begin{pmatrix} 0.0016 & 0.0003 & 0.0001 & 0 \\ 0.0025 & 1 & 0.1157 & 0 \\ 0 & 0.0002 & 0.138 & 0.0002 \\ 0 & 0.0056 & 0.0077 & 0.059 \end{pmatrix}$$

$$\bar{\boldsymbol{K}}_{d,cmax} = \begin{pmatrix} 1 & 0.25 & 0.09 & 0.34 \\ 0.41 & 0.46 & 0.72 & 0.3 \\ 0.35 & 0.1 & 0.28 & 0.52 \\ 0.51 & 0.13 & 0.13 & 0.63 \end{pmatrix}$$

The row entries in $\bar{\boldsymbol{K}}_s$ show the expected cross coupling. However, $\bar{\boldsymbol{K}}_s$ is diagonally dominant according to Equation (5.23). Hence, the decentralised controller scheme of the measurement based approach is applied. The entries of $\bar{\boldsymbol{K}}_d$ show that as for the previous examples, the dynamics are not decoupled. Hence, the control loops should not be tuned in a manner that fast transients occur, to avoid oscillations or unstable control signals. From the step responses, the controller design models are derived. Table 5.8 shows the results.

Gains	$k_{p,1}$	$k_{p,4}$	$k_{p,A}$	$k_{p,B}$
	-369	124	-44	4300
Time constants	T_1	T_4	T_A	T_B
in [s]	196	384	675	317
in cycles	0,41	0,8	1,4	0,66

Table 5.8: Model parameters for the VARICOL Ibuprofen separation

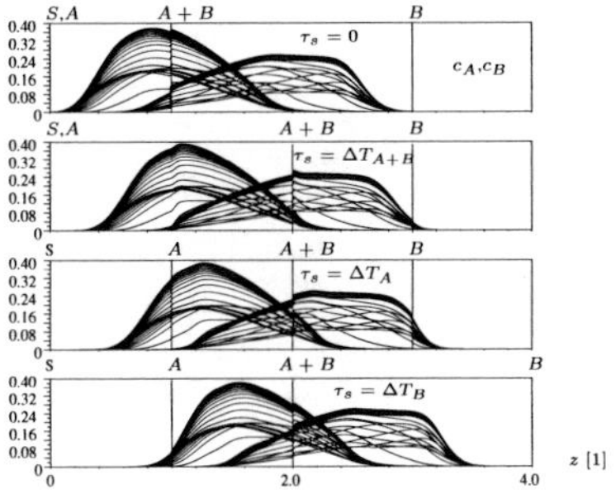

Figure 5.30: Concentration profiles during the start–up of the VARICOL plant

Gains	$k_{p,1}$	$k_{p,2}$	$k_{p,3}$	$k_{p,4}$
	$-1{,}1 \cdot 10^{-4}$	$6{,}5 \cdot 10^{-4}$	$-23 \cdot 10^{-4}$	$0{,}2 \cdot 10^{-4}$

Table 5.9: Controller gains for the VARICOL Ibuprofen separation

Controller design and closed–loop behaviour. The decentralised discrete–time PI controller scheme proposed in Section 5.2 is applied. For all control loops, $T_{w,i} = 4800\,\text{s}$ was chosen, which corresponds to a duration of 10 cycles. Table 5.9 shows the resulting controller parameters. In the following, numerical simulations of the closed–loop dynamical behaviour are presented.

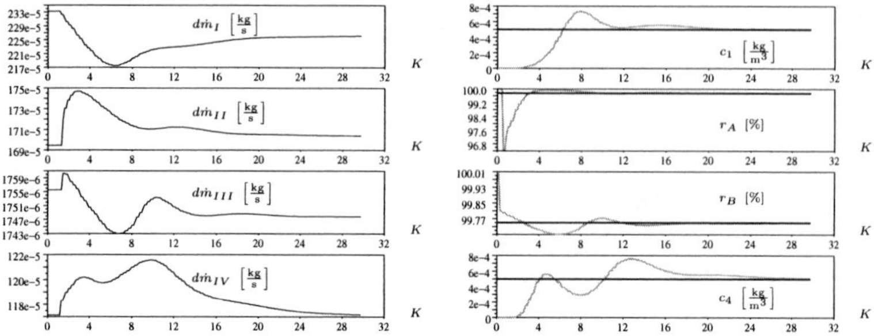

Figure 5.31: Start–up behaviour of the VARICOL plant

For the control the wave front concentrations $c_1(\frac{1}{2}T_S, z_S, k_S)$ and $c_4(\frac{1}{2}T_S, z_S, k_S)$, and the purity values $r_A(k_S)$ and $r_B(k_S)$ are considered as the controlled signals. The set–point for the wave front concentrations and the purity values is chosen as

$$\begin{aligned} c_{set,1} &= 1\cdot 5^{-4}\,\tfrac{\mathrm{kg}}{\mathrm{m}^3}\,, & c_{set,4} &= 1\cdot 5^{-4}\,\tfrac{\mathrm{kg}}{\mathrm{m}^3}\,,\\ r_{set,A} &= 99{,}75\,\%\,, & r_{set,B} &= 99{,}75\,\%\,. \end{aligned}$$

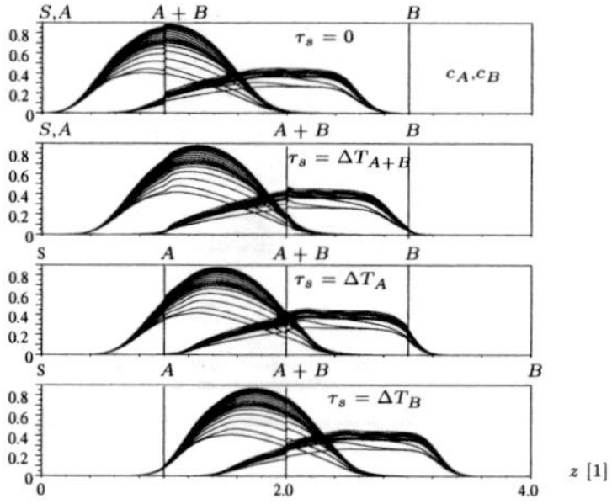

Figure 5.32: Concentration profiles during the rejection of a step disturbance of the feed inlet concentration by 100 % of the VARICOL plant

The closed–loop behaviour is investigated by numerical simulations. Figures 5.30 and 5.31 show the start–up behaviour of the VARICOL plant. Figure 5.30 shows snapshots of the concentration profiles. The left plots of Figure 5.31 show the evolution of the control input of the plant. In the right plots the evolution of the controlled signals (gray lines) and the set–point values (black lines) are shown. From the plots it can be seen that the controller drives the plant to the desired set–point. The set–points are reached after about 25 cycles after the control has been activated.

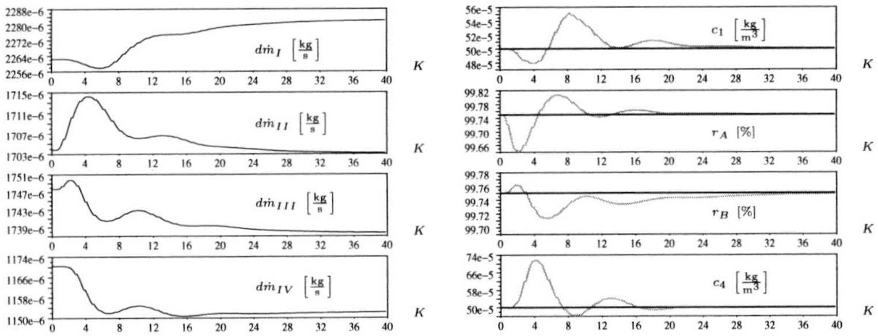

Figure 5.33: Rejection of a step disturbance of the feed inlet concentration by 100 % of the VARICOL plant

Figures 5.32 and 5.33 show the dynamical transition of the VARICOL plant during the step disturbance rejection of the feed inlet concentration by $+100\%$. The Figures show how the concentration profiles are enlarged due to the increased mass throughput. They also show that the position of the wave fronts only change very little. The set–points are reached after about 25 cycles after the step disturbance.

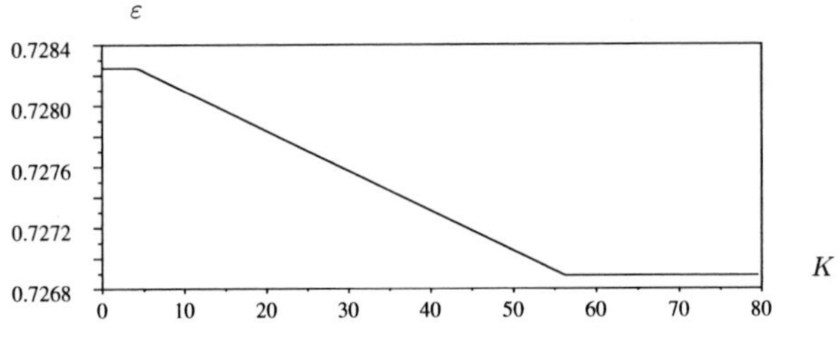

Figure 5.34: Ramp disturbance of the package porosity

The drift of the column package porosity ε is considered. Figure 5.34 shows the evolution of ε during the simulation.

Figures 5.35 and 5.36 show the transient behaviour of the VARICOL plant during the disturbance. The concentration profile plot shows that the shape and position almost remain unchanged. The left plots of Figure 5.36 show the controller input to the plant. During the ramp disturbance, the controller decreases the internal fluid flow rates. The right plots show that during the ramp disturbance a stationary control deviation occurs. After ε takes a constant value, the recycling impurity concentrations and the product purity values are met again after a transient of about 25 cycles.

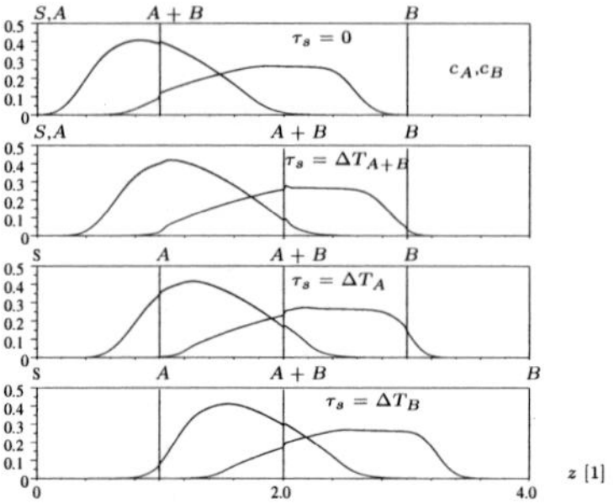

Figure 5.35: Concentration profiles during the transition of the VARICOL plant due to the ramp disturbance of the package porosity ε

5.6 Summary

In this chapter, the measurement– and observation–based continuous control of SCC processes with time–invariant switching patterns is presented. The example applications show that the decentralised discrete–time PI controllers apply well for the control of the processes. Both, the measurement–based and the observation–based controllers show a well performance with respect to driving the plant to the desired set–point and to disturbance rejection.

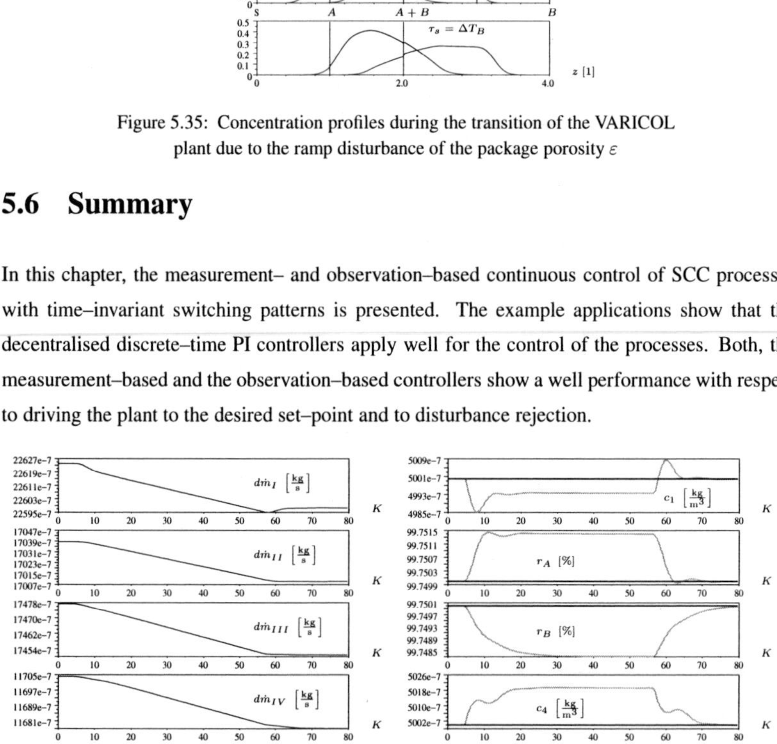

Figure 5.36: Transition of the VARICOL plant during the ramp disturbance of the package porosity ε

Solution 5.6.1 *The solution of Task 5.1.1 is given by the decentralised discrete–time controller scheme shown in Figure 5.3, if the measurement–based continuous control is considered. The controller parameters are determined by Equations (5.2) and (5.3). If the observation–based continuous control is considered, the controller scheme shown in Figure 5.6 is the solution of Task 5.1.1. The controller parameters are determined by Equations (5.2) and (5.3) for the controllers K_1 and K_4, and by Equations (5.16), (5.17), (5.20) and (5.21) for the controllers K_2, K_3, K_A and K_B.* □

Chapter 6

Combined discrete and continuous process control

Considering the port switching instants as free control input signals of SCC processes allows for the application of time–varying switching patterns by a discrete controller. The discrete controller can be combined with a continuous controller, which manipulates the internal fluid flow rates. This chapter presents an approach to the combined discrete and continuous control of SCC processes and considers the separate design of the discrete and the continuous controller.
The input signals of the discrete controller are continuously measured wave front concentrations. While keeping the internal fluid flow rates constant, the discrete controller drives the plant to a stationary operation mode, which yields a suitable switching pattern for the process. It is used as a time–invariant switching pattern for the subsequent application of a continuous controller, which drives the plant to the desired set–point.

6.1 Control task

SCC processes offer two kinds of control inputs: the discrete signals for the switching of the inlet and outlet ports, and the continuous signals of the internal fluid flow rates. The control concepts presented in Chapter 5 are based on the manipulation of the continuous control input only. It is, however, of interest to also use the discrete control input especially with respect to the implementation of SCC separations with a low number of separation columns.

For the VARICOL, the mean section length has to be predetermined for the operation of the process. In (Toumi et al., 2002b) a means for a model based parameter optimisation for the estimation of the process parameters, which are the internal fluid flow rates and the mean section lengths, is proposed. The latter determine the switching instants of the inlet and outlet ports. Because the results are applied in open–loop control, the successful application of this method strongly depends upon the exactness of the model.

In this chapter, the problem of controlling the switching instants, without predetermining the mean section length of the VARICOL process, by means of a discrete controller is addressed. Since, in general, applying a discrete control to SCC processes yields asynchronous port switching, this is referred to as the discrete control of VARICOL processes. It is not possible to drive the VARICOL to an operation point which yields the desired product purity values by a discrete controller alone. Therefore, a combination with a controller which manipulates the continuous input signals is necessary. A controller, which manipulates both the discrete and the continuous process input signals, is referred to as the *combined discrete and continuous controller* of the VARICOL.

This chapter investigates this control problem and proposes a solution for the control of following class of VARICOL processes:

Assumption 6.1.1 *For the combined discrete and continuous control of VARICOL processes with n_c separation columns, the following assumptions are made:*

1. *A time–varying switching pattern can be applied to the system by a discrete controller via the discrete control input of the plant.*
2. *Port switching can only occur at discrete times $t' = n \cdot t_s$, where $n \in \mathbb{N}^0$ and t_s is the sampling time of the discrete controller.*
3. *The internal fluid flow rates can be prescribed by a continuous controller.*
4. *An initial set of internal fluid flow rates $\dot{m}_{0,j}$, $j = I, II, III, IV$ is given.*
5. *The set–point purity values $r_{set,A}$ and $r_{set,B}$ are given.*
6. *Continuous measurements of the wave front concentrations c_i, $i = 1, 2, 3, 4$ in the column interconnections are available.*
7. *For each switching period k_S, the measurement of the product purity values $r_A(k_S)$ and $r_B(k_S)$ and the recycling impurity concentrations $c_{max,1}(z_S, k_S)$ and $c_{max,2}(z_S, k_S)$ are available.*

□

Regarding these assumptions the following **control task** is formulated:

Task 6.1.1 *The task considered for the combined discrete and continuous control of VARICOL processes is divided into the following four subtasks:*

1. *Find a switching pattern which is suitable for the initial operation point.*
2. *Drive the separation process to the desired set–point by a manipulation of the internal fluid flow rates $\dot{m}_j$, $j = I, II, III, IV$.*

3. *Reject step disturbances of the feed inlet concentrations.*

4. *Keep the plant close to the operation point in case of a ramp disturbance of the adsorbent package porosity.*

□

6.2 Way of solution

New hybrid controller design concepts propose the separated design of the discrete and the continuous part of the controller (Kamau, 2004). This concept is applied here for the solution of Task 6.1.1. For the design of the discrete controller, it is considered that the port switching has to be performed in a way which leads to low recycling and product impurity concentrations. To achieve this aim the inlet and outlet ports have to be switched by the discrete controller such that they are placed "between" the propagating wave fronts. Hence, the discrete controller is designed in form of a set of switching rules, considering continuous wave front concentration measurements according to the concept described in Section 2.5.

For the continuous controller, the same concept of decentralised discrete–time control of the recycling impurity and the product purity values as presented in Section 5.5.4 is applied. However, because the switching pattern is not known a priori, the dynamical behaviour of the process can not be determined in advance and, therefore, a modified three–step controller is applied instead of discrete–time PI controllers.

The following control concept is considered using the discrete and the continuous controller:

1. The initial internal flow rates $\dot{m}_{0,j}$, $j = I, II, III, IV$ are applied. The internal fluid flow rates are kept constant. Based on the continuous measurement of the wave front concentrations the switching rules for the discrete control of the ports are applied to start up the plant.

2. When the stationary operation mode is reached, the control loop of the discrete controller is opened, the switching pattern of the stationary operation mode is applied as a time–invariant switching pattern and the continuous control loop is closed. The separation plant is driven to the set–point and the disturbances are rejected by the continuous controller by modifying the internal fluid flow rates.

6.3 Controller design

6.3.1 Discrete control

For the design of the discrete controller the control loop shown in Figure 6.1 is considered which is based on the system representation shown in Figures 3.1 and 3.4.

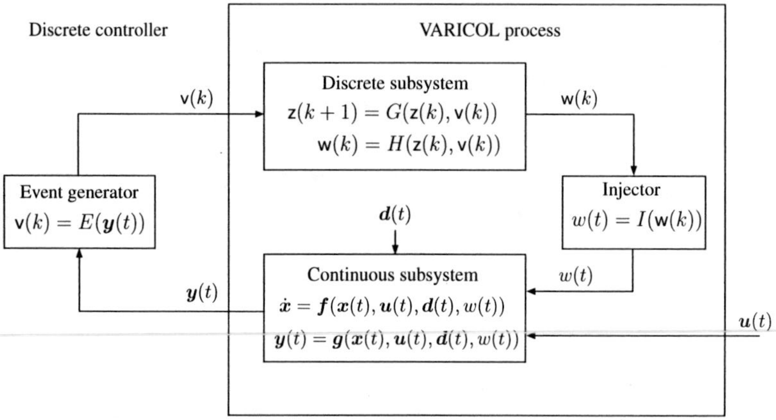

Figure 6.1: Discrete control of the VARICOL process

The event generator is considered as the discrete controller. It is specified by

$$\mathsf{v}(k) = E(\boldsymbol{y}(t))$$

and assigns to the output signal $\boldsymbol{y}$ a discrete control input v. With respect to the switching rules, which are derived to define the event generator, the continuous output is considered as

$$\boldsymbol{y}(t) = \begin{pmatrix} c_1(t, z_S + 1) \\ c_1(t, z_B + 2) \\ c_2(t, z_A + 1) \\ c_2(t, z_{A+B}) \\ c_3(t, z_{A+B} + 1) \\ c_3(t, z_B) \\ c_4(t, z_A - 1) \\ c_4(t, z_S) \end{pmatrix} .$$

The discrete control input of the VARICOL plant is

$$\mathsf{v}(k) = \begin{pmatrix} E_S(k) \\ E_A(k) \\ E_{A+B}(k) \\ E_B(k) \end{pmatrix} .$$

The continuous inputs

$$\boldsymbol{u} = \begin{pmatrix} \dot{m}_{0,I} \\ \dot{m}_{0,II} \\ \dot{m}_{0,III} \\ \dot{m}_{0,IV} \end{pmatrix}$$

and

$$\boldsymbol{d} = \begin{pmatrix} c_{A+B,A} \\ c_{A+B,B} \\ \varepsilon \end{pmatrix}$$

are considered to be constant.

In the following paragraphs, the switching rules of the event generator are derived based on the analysis of the wave front propagation with respect to the port positions.

Switching rule for the solvent inlet S. The solvent inlet S is placed between the wave fronts c_4 and c_1. Because the wave fronts perform a continuous propagation and the port can only take discrete positions, a repeated port switching has to be performed to keep the port between the two wave fronts. To decide about the switching instant, the wave front concentrations of c_1 and c_4 are continuously measured at the positions z_S and $z_S + 1$ (Figure 6.2). The following rule is considered: If the wave front concentration $c_4(t, z_S)$ is larger than the wave front concentration $c_1(t, z_S + 1)$, the solvent inlet port is switched to the next column interconnection. If the solvent inlet port is directly located next to the outlet A, the port is not switched as long as $z_A = z_S$ to guarantee the spatial sequence of the ports defined in Assumption 2.4.1. The rule is formulated in Equation (6.1).

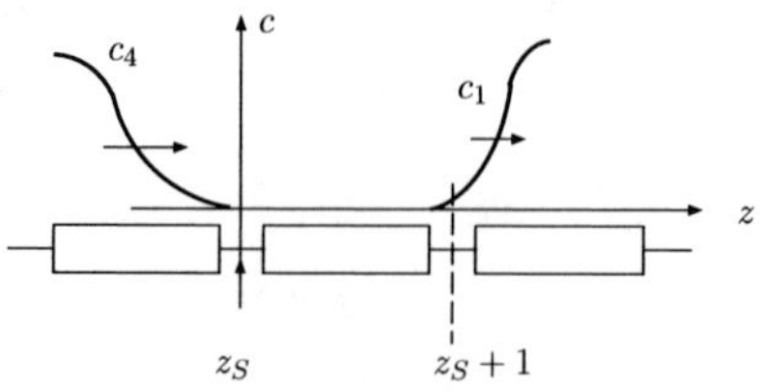

Figure 6.2: Wave front propagation at z_S

Switching rule for S :

$$\begin{aligned} &\text{if} \quad c_4(t, z_S) > c_1(t, z_S + 1) && \text{then} \quad E_S = 1 \\ & && \text{else} \quad E_S = 0\,, \\ &\text{if} \quad z_A = z_S && \text{then} \quad E_S = 0\,. \end{aligned} \tag{6.1}$$

Switching rule for the feed inlet $A + B$. The feed inlet port is placed between the wave fronts c_2 and c_3. Hence, the concentrations c_A and c_B are continuously measured at the positions z_{A+B} and $z_{A+B}+1$ to decide about the switching instant of the port (Figure 6.3). The port is switched if the wave front concentration $c_2(t, z_{A+B})$ is lower than the concentration $c_3(t, z_{A+B}+1)$ (Equation (6.2)). The feed inlet port is not switched as long as the outlet B is located at the position $z_{A+B}+1$ to guarantee that there is always one column in section III.

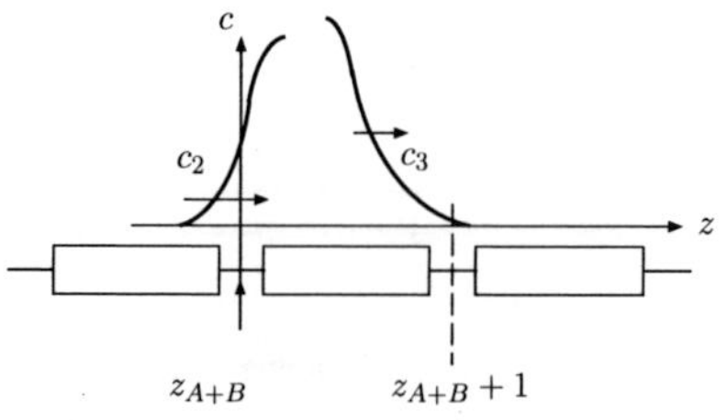

Figure 6.3: Wave front propagation at z_{A+B}

Switching rule for $A + B$:

$$\begin{aligned} &\text{if} \quad c_2(t, z_{A+B}) < c_3(t, z_{A+B} + 1) && \text{then} \quad E_{A+B} = 1 \\ & && \text{else} \quad E_{A+B} = 0\,, \\ &\text{if} \quad z_B = z_{A+B} + 1 && \text{then} \quad E_{A+B} = 0\,. \end{aligned} \tag{6.2}$$

Switching rule for the product outlet A. For the placement of the outlet A a compromise has to be found with respect to the impurity caused by the wave fronts c_4 and c_2. To decide about the switching instant of the outlet A, the wave front concentrations c_4 and c_2 are continuously measured at the positions $z_A - 1$ and $z_A + 1$ (Figure 6.4). The basic idea for the switching rule is to avoid high product impurity by the wave front c_2 and to keep the outlet port at a "secure distance" from the wave front c_4. If the port is directly located at the feed inlet position, the port is not switched. According to the switching rule formulated in Equation (6.3), the port is switched, if the wave front concentration $c_2(t, z_A + 1)$ is smaller than $\alpha\, c_4(t, z_A - 1)$, where α is a tuning factor of the switching behaviour. Limitting α to $0 < \alpha \leq 1$ guarantees that the main impurity of A is caused by the wave front c_2 and not by c_4.

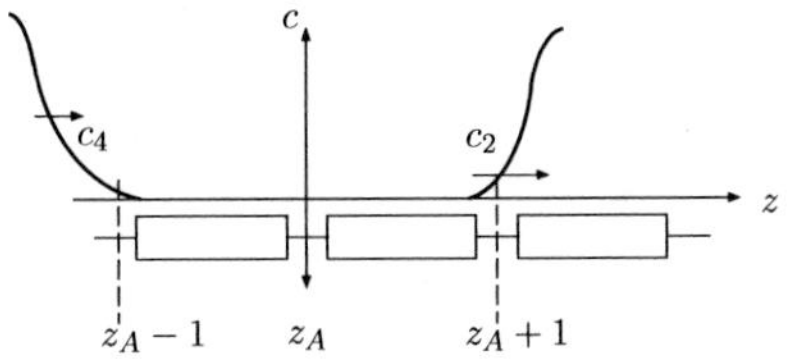

Figure 6.4: Wave front propagation at z_A

Switching rule for A :

$$\begin{aligned} &\text{if} \quad c_2(t, z_A + 1) < \alpha\, c_4(t, z_A - 1) \quad &&\text{then} \quad E_A = 1 \\ & &&\text{else} \quad E_A = 0\,, \\ &\text{if} \quad z_{A+B} = z_A &&\text{then} \quad E_A = 0\,, \\ &\text{for} \quad 0 < \alpha \leq 1\,. \end{aligned} \tag{6.3}$$

Switching rule for the product outlet B. For the switching of the outlet B, the same compromise with respect to the product impurity has to be made as was made for the outlet A. To avoid large impurities due to the by–product, the outlet has to be placed between the wave fronts c_3 and c_1. Therefore, the port switching decision is based on the continuous measurement of the wave fronts c_3 at the position z_B and of the wave front c_1 at the position $z_B + 2$ (Figure 6.5). The port is switched, if $c_3(t, z_B)$ is larger than $\beta\, c_1(t, z_B + 2)$, where β is a tuning factor. The port is not switched, if it is located directly next to the solvent inlet. The switching rule is formulated in Equation (6.4). If β is limited to $0 < \beta \leq 1$ it is guaranteed that the main impurity in the product outlet B is caused by the wave front c_3 and not by c_1.

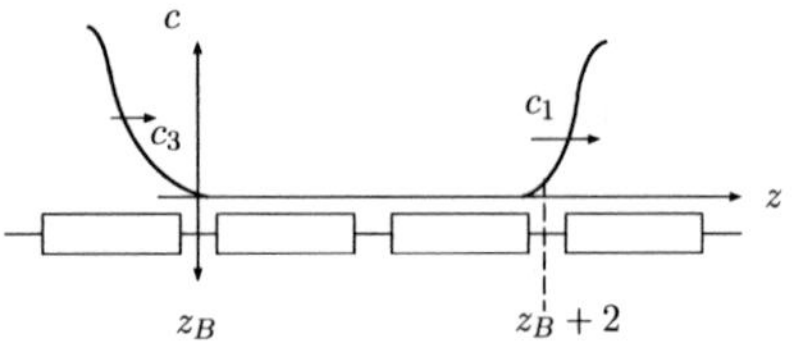

Figure 6.5: Wave front propagation at z_B

Switching rule for B :

$$
\begin{array}{llll}
\text{if} & c_3(t, z_B) > \beta\, c_1(t, z_B + 2) & \text{then} & E_B = 1 \\
 & & \text{else} & E_B = 0\,, \\
\text{if} & z_S + n_c = z_B & \text{then} & E_B = 0\,,
\end{array}
\qquad (6.4)
$$

for $0 < \beta \leq 1\,.$

Event generator. The event generator E is defined by the switching rules (6.1) through (6.4). Applying the event generator for the discrete control of the VARICOL yields a switching pattern, which is time–varying in the transient *and* in the stationary operation mode of the VARICOL because port switching is not possible at arbitrary time points, but only at the sampling instants of the event generator at the discrete times $t' = n\, t_s$. According to the choice of α and β, the switching pattern yields a unique purity and recycling impurity concentration evolution in the stationary operation mode for given continuous input signals $\boldsymbol{u}$ and $\boldsymbol{d}$.

A sampling of the continuous plant output $\boldsymbol{y}(t)$ is performed at discrete times $t' = n \cdot t_s$, $n \in \mathbb{N}^0$. At the sampling instant, the rules of the event generator are checked and the resulting new port switching signal is applied to the VARICOL process.

Because each port is regarded separately and the corresponding wave front concentration measurements are considered with respect to the port position, this concept applies to VARICOL processes with an arbitrary number of columns $n_c > 3$. It has been tested in numerical simulations for processes with 4, 5, 6 and 8 columns. It is theoretically possible to apply the concept to a 3–column process.

6.3.2 Continuous control

For an unfavourable choice of the continuous input signals a discrete control alone can not yield the desired product purity values. To fulfil this aim, a continuous–variable controller has to be applied, which manipulates the internal fluid flow rates such that the desired product purity values are obtained.

The continuous controller, which is proposed here, is applied considering a time–invariant switching pattern. Therefore, the control concept presented in Section 5.5.4 can be used. The controlled process output variables are the wave front concentrations $c_{m,1}(k_S)$ and $c_{m,4}(k_S)$ in the recycling and the purity values $r_A(k_S)$ and $r_B(k_S)$, which all are measured and provided to the controller in each switching period k_S. Because the wave front concentrations c_1 and c_4, which occur in the moment of the switching of port S, are the largest impurity concentrations $c_{max,1}(z_S, k_S)$ and $c_{max,4}(k_S)$, these values are chosen as the controlled variables

$$\begin{aligned} c_{m,1}(k_S) &= c_{max,1}(z_S, k_S) \\ c_{m,4}(k_S) &= c_{max,4}(z_S, k_S) \,. \end{aligned}$$

It is assumed, that the cross coupling of the process is low. Therefore, a controller scheme as shown in Figure 6.6 with the decentralised controllers $K_j, j = I, II, III, IV$ is applied.

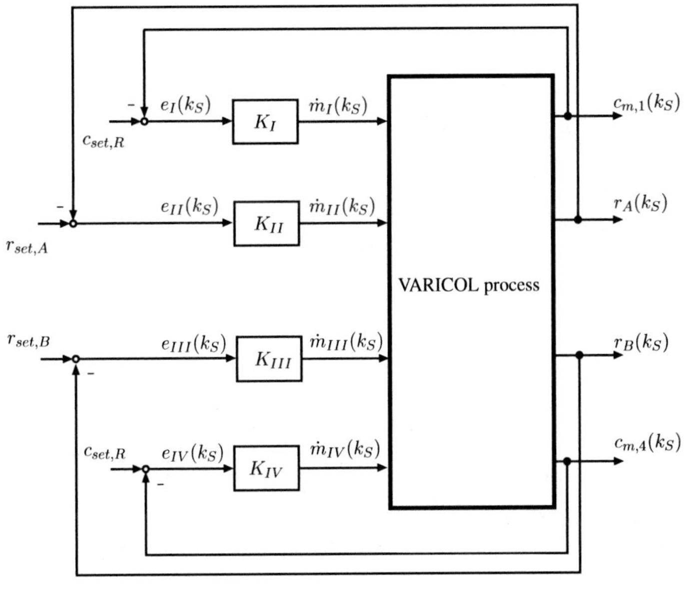

Figure 6.6: Continuous control of the VARICOL process

The dynamics of the input–output behaviour can not be predetermined because the switching pattern is not known in advance. Hence, modified discrete–time three–step controllers, which are sampled in the moment of the switching of the port S, are used instead of decentralised PI controllers. A modified three–step controller is shown in Figure 6.7. The control deviation $e_j(k_S)$ is converted to the step signal $u_j(k_S)$, $j = I, II, III, IV$, which is summed up when the sampling of the system takes place. The tuning parameters of the controller are the range δe_j for the control deviation and the step size δu_j. The control law is given by

$$\begin{aligned} &\text{if} \quad |e_j(k_S)| \geq \delta e_j\,, \quad \text{then} \quad u_j(k_S) = \text{sign}(e_j(k_S))\,\delta u_j\,, \\ &\text{else} \qquad\qquad\qquad\qquad\quad u_j(k_S) = 0\,, \\ &\qquad\qquad \dot{m}_j(k_S) = \dot{m}_j(k_S - 1) + u_j(k_S)\,. \end{aligned} \tag{6.5}$$

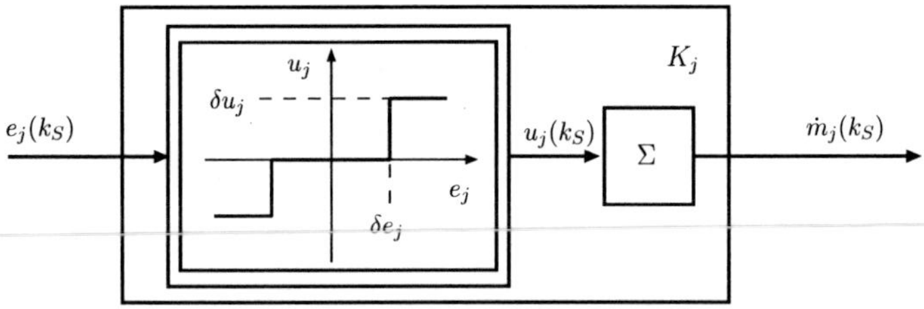

Figure 6.7: Modified three–step controller

The modified three–step controller has the property that as long as the control deviation lies outside of the interval $\pm\delta e_j$ (Figure 6.8), a successive change of the controller output is performed. If the control deviation lies within the interval $\pm\delta e_j$, the controller output remains unchanged.

For the implementation of the controller scheme, the parameters δe_j and δu_j have to be chosen in advance. For the interval δe_j, the following values are chosen:

- $\delta e_I = 10\,\%$,
- $\delta e_{II} = 0{,}5\,\%$,
- $\delta e_{III} = 0{,}5\,\%$,
- $\delta e_{IV} = 10\,\%$.

The values δu_j can be chosen based on the triangle theory (Migliorini et al., 1998). According to the triangle theory, an interval $\Delta\dot{m}_{II}$ and $\Delta\dot{m}_{III}$ is determined for $\dot{m}_{II}$ and $\dot{m}_{III}$, for which a $100\,\%$ product purity is obtained. A further interval $\Delta\dot{m}_I$ and $\Delta\dot{m}_{IV}$ is determined for $\dot{m}_I$ and $\dot{m}_{IV}$, for which the wave fronts c_1 and c_4 are kept within the sections I and IV. The step sizes δu_j are determined from these intervals by

$$\delta u_j = f_{u,j}\,\Delta\dot{m}_j\,,$$

with $f_{u,j} \in [\frac{1}{1000}, \frac{1}{500}]$. The interval was determined heuristically based on the results of several simulations.

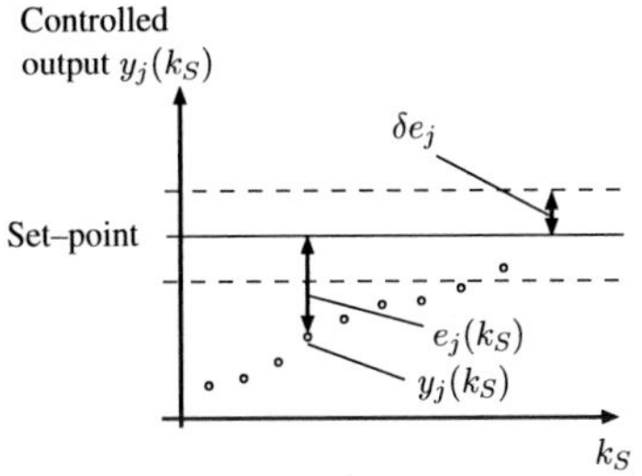

Figure 6.8: Set–point interval of the modified three–step controller

6.4 Application example

The performance of the combined discrete and continuous control is tested with numerical simulations of an example VARICOL process. For the process setup and the separation example, the same system as in Section 5.5.4 is used. The same initial operation point is applied.

The following controller parameters where used:

- $t_S = 5\,\mathrm{s}$ Sampling time of the discrete controller,
- $\alpha = 0.02$ Tuning parameter for the product port switching,
- $\beta = 0.03$ Tuning parameter for the product port switching,
- $r_{set,A} = 95\,\%$ Purity set–point for the product A,
- $r_{set,B} = 95\,\%$ Purity set–point for the product B,
- $\delta u_I = 5 \cdot 10^{-7} \frac{\mathrm{kg}}{\mathrm{s}}$ controller output step width for K_I,
- $\delta u_{II} = 5 \cdot 10^{-7} \frac{\mathrm{kg}}{\mathrm{s}}$ controller output step width for K_{II},
- $\delta u_{III} = 2 \cdot 10^{-7} \frac{\mathrm{kg}}{\mathrm{s}}$ controller output step width for K_{III},
- $\delta u_{IV} = 5 \cdot 10^{-7} \frac{\mathrm{kg}}{\mathrm{s}}$ controller output step width for K_{IV}.

The set–point values $c_{set,I}$ and $c_{cset,IV}$ are determined from the recycling impurity concentrations $c_{m,1}$ and $c_{m,4}$ in the stationary operation mode of the discretely controlled plant.

The following steps of the simulation are performed:

1. The plant is started applying the initial internal fluid flow rates. The discrete control is applied and the switching is performed according to the switching rules (6.1) through (6.4).

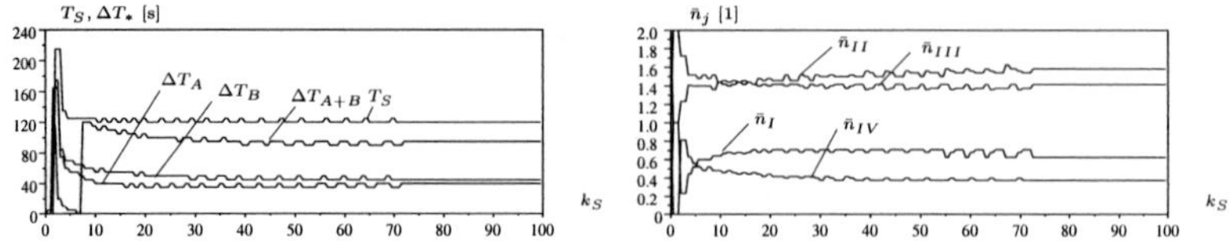

Figure 6.9: Switching times and mean section lengths during the discretely controlled start–up of the VARICOL plant

The internal fluid flow rates are kept constant and it is assumed that no variation of the disturbance input occurs.

2. After the plant has reached a stationary operation mode, the resulting switching pattern is determined, the discrete control loop is opened and the switching pattern is applied as a time–invariant pattern. At the same time, the mean values of the recycling impurity concentrations $c_{m,1}$ and $c_{m,4}$ are determined and applied as the set–point values for the controllers K_I and K_{IV}. The continuous control is applied and the plant is driven to the desired set–point.

3. After the set–point is reached, a disturbance step input of $+25\,\%$ is applied to the feed inlet concentrations. The transient behaviour of the continuous control loop during the disturbance rejection is recorded.

4. A separate numerical simulation is performed to test the closed–loop behaviour in case of a drift of the adsorbent package porosity.

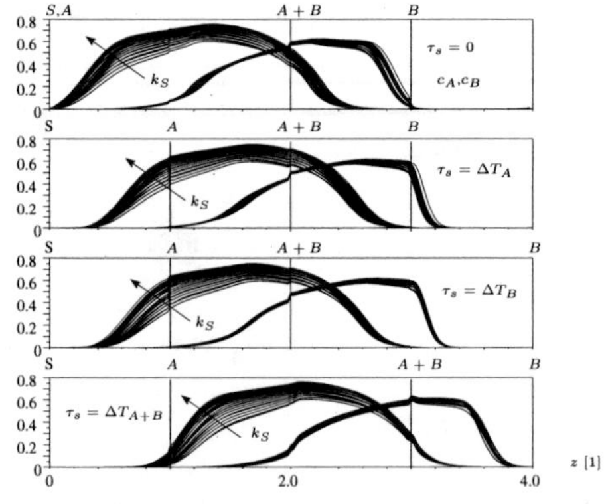

Figure 6.10: Snapshots of the concentration profiles during the discretely controlled start–up after $k_S = 7$

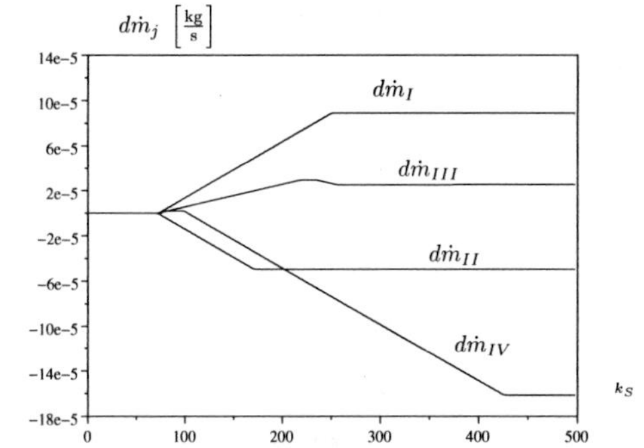

Figure 6.11: Control input during the discretely controlled start–up and after activating the continuous control at $k_S = 74$

Figure 6.9 shows how the switching times and the mean section lengths evolve during the discretely controlled start–up of the VARICOL plant. The left plot shows the signals of T_S and the relative switching times ΔT_A, ΔT_{A+B} and ΔT_B, and the right plot shows $\bar{n}_{c,I}$, $\bar{n}_{c,II}$, $\bar{n}_{c,III}$ and $\bar{n}_{c,IV}$. The left plot shows strong changes of the switching times during the first five switching periods k_S. Then, a clear tendency of the switching times establishes.

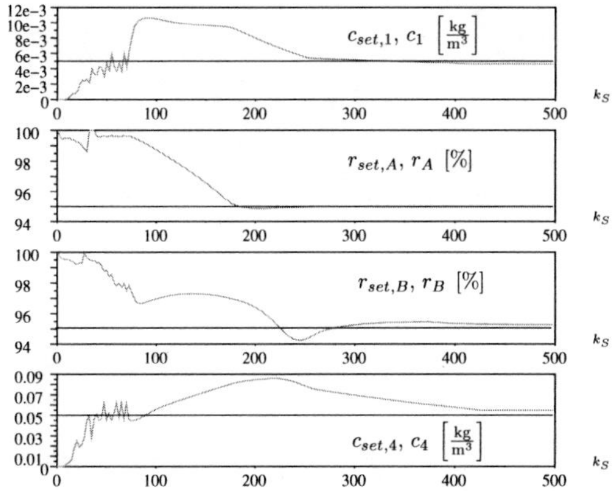

Figure 6.12: Continuous plant output and set–point values during the discretely controlled start–up and after the activation of the continuous control at $k_S = 74$

At $k_S = 7$, the switching time ΔT_{A+B} performs a considerable step change, which results in a change of the port switching sequence. After this change, the switching times remain almost unchanged and perform a slight transition to a stationary mode. In other words, from $k_S = 7$ on, the cyclic sequence of port switchings remains unchanged. The right plot shows how the mean section lengths evolve.

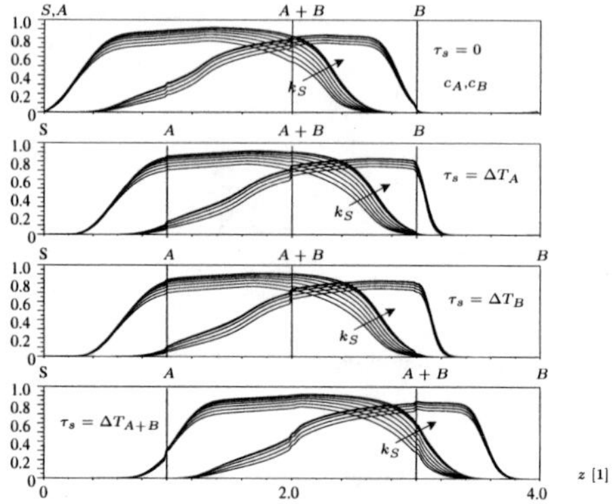

Figure 6.13: Snapshots of the concentration profiles after the continuous control is activated at $k_S = 74$

A similar signal trajectory compared to the switching times can be observed. After the first five switching periods the signals show a clear tendency with respect to the convergence to a stationary value. At $k_S = 7$ the mean section length $\bar{n}_{c,I}$ becomes larger than $\bar{n}_{c,IV}$, which is a result from the change of the relative switching time ΔT_{A+B} at the same time. After $k_S = 7$, the mean section lengths converge to a stationary mode. The periodic property of the signal trajectories for $k_S > 55$ is caused by the fact that switching is only possible at prescribed times $t' = n\,t_S$ and not at arbitrary times t'. At $k_S = 74$, the discrete control loop is opened and the switching pattern is kept time–invariant. Therefore, the signals take constant values from this moment on.

Figure 6.10 shows the snapshots of the concentration profiles during the discretely controlled start–up after $k_S = 7$. The plots show how the concentration profiles evolve and converge to a stationary mode.

Figure 6.11 shows the control input to the varicol plant during the start–up. It shows that the signals are changed by the controller after the continuous control is applied at $k_S = 74$. The plot shows how the controller adapts the internal flow rates to drive the controlled variables to the set–point.

The trajectories of the controlled variables during the start–up are shown in Figure 6.12. The plots show how the wave front concentrations $c_{m,1}(k_S)$ and $c_{m,4}(k_S)$ and the purity values $r_A(k_S)$ and $r_B(k_S)$ evolve during the discretely controlled start–up from $k_S = 0$ to $k_S = 74$. It can be seen

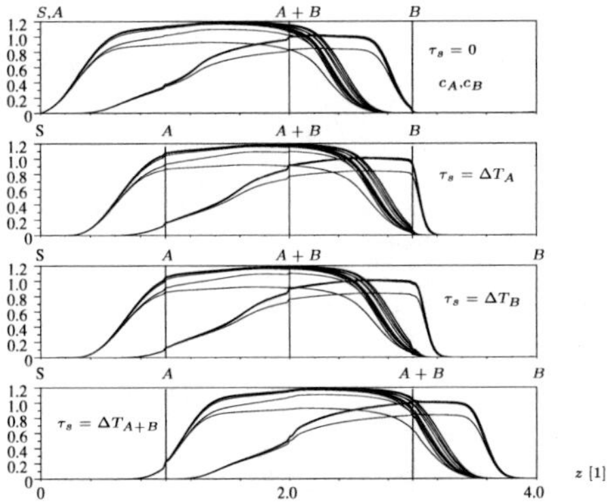

Figure 6.14: Snapshots of the concentration profiles after the step disturbance of the feed inlet concentration

that in the last periods before activating the continuous control, the transient of the signals during the discretely controlled start–up is finished. Furthermore, the plots show how the variables are driven close to their set–point under the influence of the modified three–step controllers after the continuous control has been activated. The continuous transition is terminated after about 300 periods, which corresponds to 75 cycles. This transition is slow compared to the transition obtained with the continuous controllers proposed in Chapter 5. However, the controller scheme is simple and easy to initialise. With respect to the unknown dynamical behaviour, the controller parameters are tuned such that the transition is slow but stable. Figure 6.13 shows the snapshots of the concentration profiles during the continuously controlled start–up of the plant.

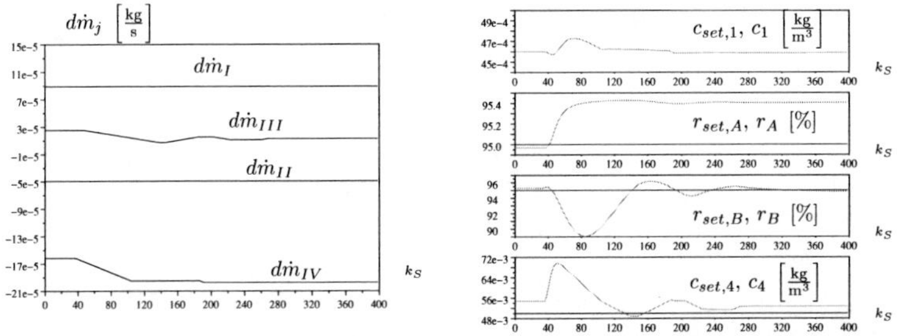

Figure 6.15: Continuous control input of the VARICOL process and controlled signals during the disturbance rejection of the feed inlet concentration

After the stationary state has been reached, a step disturbance is applied to the feed inlet concentration by $+25\%$ while keeping the plant under continuous control. Figure 6.14 shows the snapshots of the concentration profiles after the step disturbance has been applied. Figure 6.15 shows how the continuous control input and the controlled signals evolve. The plots show that the continuous controller can reject the disturbance.

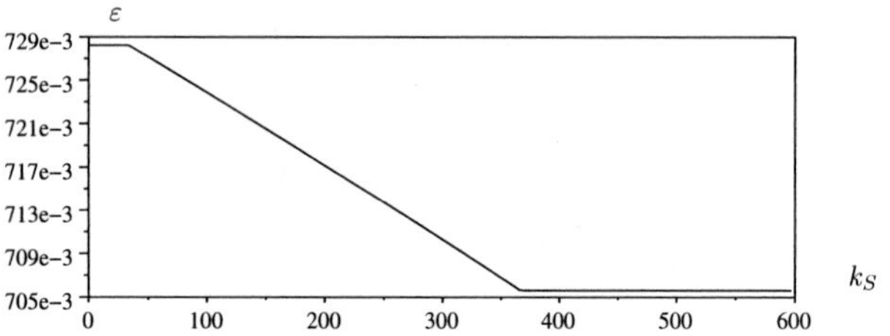

Figure 6.16: Package porosity drift during the simulation

In a further simulation, a drift of the package porosity was assumed to occur after the continuous controller has been activated and the stationary state is reached. Figure 6.16 shows the drift of the package porosity which was assumed for the simulation.

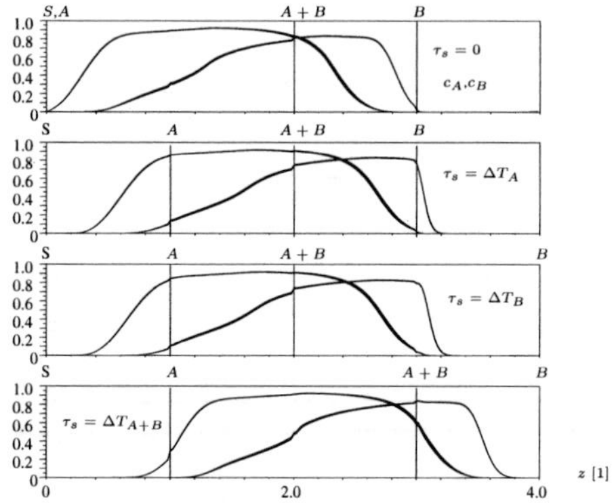

Figure 6.17: Snapshots of the concentration profiles during the drift of the package porosity

Figure 6.17 shows the snapshots of the concentration profiles during the drift of the package porosity. It shows that the shape changes marginally. Figure 6.18 shows, how the control input is manipulated and how the controlled variables evolve during the disturbance by the package drift.

The simulation studies show how the controller handles the start–up of the plant and the rejection of the feed inlet concentration and the package porosity disturbance. The examples show that the control of VARICOL processes based on the manipulation of the discrete *and* the continuous

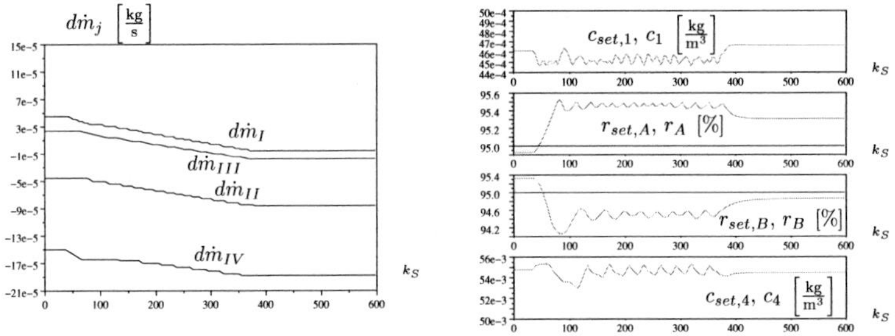

Figure 6.18: Continuous control input of the VARICOL process and controlled signals during and after the drift of the package porosity

input signals can be performed by the sequential application of the discrete and the continuous controllers.

6.5 Summary

The discrete controller for VARICOL processes presented in this chapter allows for the discrete control of the port switching based on the continuous measurement of the wave front concentrations in selected column interconnections. The controller is implemented as the event generator of the hybrid SCC system. The switching rules are easy to implement and can be tuned by the adaptation of the parameters α and β during process operation. The switching pattern, which is obtained in the stationary state from the application of the discrete controller, is a suitable time–invariant pattern for the application of the continuous controller in form of the modified three–step controller to drive the plant to the desired set–point.

Both the discrete and the continuous controllers are based on simple control laws. The numerical simulations show that the concept gives satisfactory results for the given example. The presented approach is not restricted to SCC processes with four separation columns.

Solution 6.5.1 *A solution to Task 6.1.1 is given by the sequential application of the discrete controller of VARICOL processes, which is the event generator given by Equations (6.1) through (6.4), and the decentralised modified three–step controller given by Equation (6.5).* □

Chapter 7

Conclusion

Summary. The aim of this thesis is to contribute to the solution of the controller design problem for Simulated Counterflow Chromatographic (SCC) separations by presenting and discussing new control concepts, which provide simple and easy to handle state observation and control algorithms. The central aim is the control of the product purity. Providing a detailed process description and an elaborate physical modelling in Chapter 2 and 3, the thesis gives the basis for the analysis of the SCC process with respect to the control aim. The main emphasis is made on the representation of the process behaviour and the options for the manipulation of the product purity. A new representation concept of the wave fronts of the SCC concentration profiles is described. The manipulation of the wave fronts is a key issue with respect to the control of both SMB and VARICOL processes. For the description of the port switching, the concept of time–invariant and time–varying switching patterns is introduced, which is used for the modelling of the discrete part of the processes.

Based on the process description, a physical model is derived in Chapter 3 with the aim to capture the continuous–variable and the discrete–event dynamics of SCC processes as well as their interaction, to provide a basis for an analytical investigation of the process behaviour. On the one hand, the modelling provides the basis for the implementation of a numerical plant simulation. On the other hand, the model allows to study the continuous and the discrete dynamics separately, and reveals the properties of the combined discrete and continuous dynamics. The continuous subsystem is modelled with a physical model of each separation columns. The coupled column models are represented in a block diagram to show up the model structure. Based on this representation, the impact of the discrete subsystem onto the continuous system is studied. The discrete subsystem is modelled by means of a finite deterministic automaton. The discrete state, input and output variable are introduced to describe the dynamics of the port switching, which are represented by a state transition relation and an output relation of a finite deterministic automaton. The results are applied for the modelling of SCC process encountering synchronous and asynchronous port switching as well as time–invariant and time–varying switching patterns. It is found that the discrete model shows differnet degrees of complexity for different process setups. The simplest dynamical behaviour is obtained for an SMB process with a moving spatial coordinate system representation of the component concentration profiles. The most complex discrete dynamics are obtained for VARICOL processes with a time–varying switching pattern and a fixed spatial coordinate system. The overall system dynamics of the coupled continuous and discrete models is represented using the concept of hybrid automata. The representation is applied to processes with a time–invariant switching pattern. For different process configurations either jumps of the continuous state, or switching continuous dynamics, or both, occur. By the separate modelling of the subsystems, and the modelling of the overall system dynamics, it is shown that the coupling of the continuous output and the discrete input is not prescribed by the physical plant set–up and, hence, has to be designed for a successful operation of the plant.

For the control, either the direct measurement or the reconstruction of the concentration profiles is necessary. Because of the high complexity of SCC processes and the low number of available measurements, a simple approach is chosen to derive a reconstruction of the concentration profile wave fronts in Chapter 4. Based on the explicit stationary solution of the physical true counterflow model, a functional description of the form and propagation of the SMB wave fronts is derived. The parameters of the description are unique for each wave front for the actual operation mode. It is shown how the parameters can be determined by a linear discrete–time observer. The resulting observation algorithm shows low complexity and the design of the observer is considerably easy. Except of the switching time of the SMB and the measurement times of the wave front concentration measurements, no parameters are needed. The parameters of the physical model are implicitly contained in the states of the wave front model, which are determined by the observation. On the basis of an observation error analysis, a rule for the choice of the measurement times is derived. The discussion of numerical simulation results of the observation of a controlled SMB plant shows the observer performance under practically relevant conditions.

Despite of the complex dynamics, the control of SCC processes, to which a time–invariant switching pattern is applied, can be performed using standard decentralised discrete–time PI controllers. Two concepts are analysed in detail in Chapter 5: the measurement–based control scheme consisting of four standard control loops, and the observation–based control scheme, which uses decentralised cascaded control loops. The inner loops control a wave front parameter, which is determined by the wave front form and position, and the outer loops control the product purity values. For the controller design, the process dynamics have to be identified from numerical simulations or experimental tests. Similar control concepts have allready been published in literature. The novelty of the herein presented approach consists of the direct measurement of the product purity values and the use of the wave front observation presented in Chapter 4 within the control scheme. The results are presented in form of control algorithms of low complexity, of which the design follows a simple procedure, and in form of numerical simulations of the control of SMB and VARICOL processes.

The variation of the switching pattern offers a further degree of freedom for the control of SCC processes. In Chapter 6, the first known approach to discrete closed–loop control and combined discrete and continuous control of SCC processes is described. Because of the complexity of the continuous dynamics the methods of hybrid systems theory do not apply for the design of a controller. As it is the main objective to derive controllers of a simple structure, a strongly simplified description of the process behaviour considering only the propagation of the four wave fronts with respect to the inlet and outlet port positions is used to design the discrete controller. A set of rules, which determine the switching instants of each port based on continuous wave front concentration measurements, is derived and tested in numerical simulations of the process. It is found that the principle can be applied to control the start–up of the plant. If the continuous plant input signals are constant, a cyclic stationary state is reached, which yields a suitable switching pattern for the considered VARICOL process. To drive the plant to the desired set–point, the stationary switching pattern is applied as a time–invariant pattern and a continuous controller is applied. The closed–loop behaviour is investigated based on numerical simulation results.

Outlook. Possible extentions of this work can be categorised into two fields. One field regards open problems with respect to the control of SCC processes. Because of the simplicity of the observation and control algorithms presented in this thesis it is worth applying the results to real plants. Furthermore, a combination of the presented approach with different operation concepts like the Modicon or the Powerfeed (Kloppenburg and Gilles, 1998; Schramm et al., 2002b), or the application of the wave front reconstruction to the VARICOL principle with time–invariant and time–varying switching patterns can be considered. The most challenging task seems to be the parallel application of the discrete and the continuous control of SCC processes. A possible approach is to prescribe both the internal fluid flow rates and the mean section length by a continuous controller.

The second field regards SCC processes from the system theoretical point of view. The processes generally exhibit the combined discrete and continuous dynamics, because the SCC is based on the combination of continuous and discrete actions. As could be shown in Chapter 3, the processes show different switching phenomena under different operation conditions. The application of existing, or the development of new suitable controller design methods for these kind of systems are open problems. However, because the continuous subsystem is a distributed–

parameter system, these problems are difficult to solve. The transformation of the continuous model to a lumped–parameter continuous–time system opens up several possibilities with respect to the design of continuous and discrete controllers. It allows to study several hybrid phenomena on a practically relevant process. However, it has to be regarded that a transformation of the continuous model to a lumped–parameter system yields a continuous subsystem model of large dimension, which can render the solution of this problems impossible. Nevertheless, the phenomena, which an SCC process with time–invariant and time–varying switching patterns reveals, offer a broad field for the analysis and the design of hybrid dynamical systems.

Bibliography

S. Abel, G. Erdem, M. Amanullah, M. Morari, M. Mazzotti, and M. Morbidelli. Optimizing control of simulated moving beds - experimental implementation. *Journal of Chromatography A*, 1092(1):2–16, 2005.

S. Abel, G. Erdem, M. Mazzotti, M. Morari, and M. Morbidelli. Model predictive control of simulated moving bed separations. In *Book of abstracts of the Symposium on Preparative and Industrial Chromatography and Allied Techniques*, Heidelberg, Germany, 2002.

R. Alur, C. Courcoubetis, T. Henzinger, and P. Ho. *Hybrid Automata: An algorithmic approach to the specification and verification of hybrid systems*, volume 736 of *Lecture Notes in Computer Science*. Springer–Verlag, Berlin, 1993.

D. Broughton and C. Gerhold. Continuous sorption process employing fixed bed of sorbent and moving inlets and outlets. *US Patent 2985589*, 1961.

P. V. Danckwerts. Continuous flow systems. *Chemical Engineering Science*, 68(3):1–13, 1953.

A. Depta. *Präparative Gegenstromchromatographie mit überkritischem Kohlendioxid*. Dissertation, Technische Universität Hamburg–Harburg, 1999.

A. Depta, T. Giese, M. Johannsen, and G. Brunner. Separation of stereoisomers in a simulated moving bed - supercritical fluid chromatography plant. *Journal of Chromatography A*, 865: 175–186, 1999.

G. Dünnebier and K.-U. Klatt. Modelling and simulation of nonlinear chromatographic separation processes: a comparison of different modelling approaches. *Chemical Engineering Science*, 55:373–380, 2000.

G. Dünnebier, I. Weirich, and K.-U. Klatt. Computationally efficient dynamic modelling and simulation of simulated moving bed chromatographic processes with linear isotherms. *Chemical Engineering Science*, 53(14):2537–2546, 1998.

G. Erdem, S. Abel, M. Mazzotti, M. Morari, M. Morbidelli, and J. H. Lee. Automatic Control of Simulated Moving Beds. *Industrial and Engineering Chemistry Research*, 43(2):405–421, 2004.

T. Giese. *Simulation der Chromatographie mit überkritischem Kohlendioxid am Beispiel der Trennung eines Diterpens*. Dissertation, Technische Universität Hamburg–Harburg, 2002.

S. Golshan-Shirazi and G. Guiochon. Comparison of the various kinetic models of non–linear chromatography. *Journal of Chromatography A*, 603:1–11, 1992.

F. Hanisch. *Prozessfürung präparativer Chromatographieverfahren*. Dissertation, Universität Dortmund, 2002.

M. Johannsen. *Präparative Chromatographie mit überkritischen Gasen*. Mensch & Buch Verlag, Berlin, 2004.

S. I. Kamau. *Modelling, Analysis and Design of Discretely Controlled Switched Positive Systems*. Dissertation, Ruhr–Universität Bochum, 2004.

A. Kienle. *Nichtlineare Wellenphänomene und Stabilität stationärer Zustände in Destillationskolonnen*. Dissertation, Universität Stuttgart, 1997.

K.-U. Klatt. Modellierung und effektive numerische Simulation von chromatographischen Trennprozessen im SMB-Betrieb. *Chemie Ingenieur Technik*, 71(6):555, 1999.

K.-U. Klatt, F. Hanisch, and G. Dünnebier. Model-based control of a Simulated Moving Bed chromatographic process for the separation of fructose and glucose. *Journal of Process Control*, 12(2):203–219, 2002.

K.-U. Klatt, F. Hanisch, G. Dünnebier, and S. Engell. Model–based optimization and control of chromatographic processes. In *Proceedings of the 7th int. Symposium on Process Systems Engineering*, Keystone, USA, 2000.

T. Kleinert. Theoretical modelling and dynamic simulation of simulated counterflow chromatographic separations. Research report, Arbeitsbereich Regelungstechnik,Technische Universität Hamburg–Harburg, 2002.

T. Kleinert and J. Lunze. A hybrid automaton representation of simulated counterflow chromatographic separation processes. In *Proceedings of the 15. IFAC World Congress*, Barcelona, Spain, 2002.

T. Kleinert and J. Lunze. Modelling and state observation of Simulated Moving Bed processes. In *Proceedings of the European Control Conference ECC 2003*, Cambridge, England, 2003.

T. Kleinert and J. Lunze. Modellierung und Zustandsbeobachtung von Simulated-Moving-Bed Prozessen. *Automatisierungstechnik*, Volume 42(11):503–513, 2004.

T. Kleinert and J. Lunze. Modelling and and state observation of Simulated Moving Bed processes based on an explicit functional wave form description. *Mathematics and Computers in Simulation*, Volume 68(3):235–270, 2005.

E. Kloppenburg. *Modellbasierte Prozessführung von Chromatographieprozessen mit simuliertem Gegenstrom*. Dissertation, Universität Stuttgart, 2000.

E. Kloppenburg and E. Gilles. Ein neues Prozeßführungskonzept für die Chromatographie mit simuliertem Gegenstrom. *Chemie Ingenieur Technik*, 70(12), 1998.

E. Kloppenburg and E. Gilles. Automatic control of the simulated moving bed process for C8 aromatics separation using asymptotically exact input/output-linearization. *Journal of Process Control*, 9(1):41–50, 1999.

L. Lapidus and N. R. Amundson. The effect of longitudinal diffusion in ion exchange and chromatographic columns. *Journal of Physical Chemistry*, 56:984–988, 1952.

M. Lübbert. *Adsorption aus überkritischen Lösungen*. Dissertation, Technische Universität Hamburg–Harburg, 2004.

O. Ludemann-Homburger, R. Nicoud, and M. Baley. The varicol process: a new multicolumn continuous chromatographic process. *Separation Science and Technology*, 35(12):1829–1862, 2000.

G. Ludyk. *Theorie dynamischer Systeme*. Elitera–Verlag, 1977.

G. Ludyk. *Time–Variant Discrete–Time Systems*. Vieweg–Verlag, 1981.

J. Lunze. *Robust Multivariable Feedback Control*. Prentice Hall, 1989.

J. Lunze. What is a hybrid system? In S. Engell, G. Frehse, and E. Schnieder, editors, *Modelling, Analysis and Design of Hybrid Systems*, volume 279 of *Lecture notes in control and information sciences*, pages 3–14. Springer-Verlag, Berlin, 2002.

J. Lunze. *Automatisierungstechnik*. Oldenbourg Verlag, 2003.

J. Lunze. *Regelungstechnik 1*. Springer–Verlag, 2004a.

J. Lunze. *Regelungstechnik 2*. Springer–Verlag, 2004b.

M. Mangold, G. Lauschke, J. Schaffner, M. Zeitz, and E.-D. Gilles. State and parameter estimation for adsorption columns by nonlinear distributed parameter state observers. *Journal of Process Control*, 4(3):163–172, 1994.

W. Marquardt. *Nichtlineare Wellenausbreitung - ein Weg zu reduzierten Modellen von Stofftrennprozessen*. Dissertation, Universität Stuttgart, 1988.

W. Marquardt. Traveling waves in chemical processes. *International Chemical Engineering*, 30 (4):585–506, 1990.

C. Migliorini, M. Mazzotti, and M. Morbidelli. Continuous chromatographic separation through simulated moving beds under linear and nonlinear conditions. *Journal of Chromatography A*, 827:161–173, 1998.

S. Natarajan and J. H. Lee. Repetitive model predictive control applied to a simulated moving bed chromatography system. *Computers and Chemical Engineering*, 24:1127–1133, 2000.

S. Peper, S. Crammerer, M. Johannsen, and G. Brunner. Supercritical fluid chromatography: Process optimization of the separation of tocopherol homologues. In *Proceedings of the International Symposium on Supercritical Fluids*, Versailles (France), 2003.

S. Peper, M. Lübbert, M. Johannsen, and G. Brunner. Separation of ibuprofen enantiomers by supercritical fluid simulated moving bed chromatography. *Separation Science and Technology*, 11(37):2545–2566, 2002.

F. Raske. *Ein Beitrag zur Dekomposition von linearen zeitinvarianten Grossystemen*. Dissertation, Universität Hannover, 1981.

D. Ruthven and C. Ching. Counter-current and simulated counter-current adsorption separation processes. *Chemical Engineering Science*, 44(5):1011–1038, 1998.

H. Schramm, S. Grüner, and A. Kienle. Optimal operation of simulated moving bed chromatographic processes by means of simple feedback control. *Journal of Chromatography A*, 1006: 3–13, 2003.

H. Schramm, S. Grüner, A. Kienle, and E. Gilles. Control of moving bed chromatographic processes. In *Proceedings of the European Control Conference ECC 2001*, Porto, Portugal, 2001.

H. Schramm, S. Grüner, A. Kienle, and E. D. Gilles. Improving simulated moving bed processes by cyclic modulation of the feed concentrations. *Chemical Engineering Technology*, 25:1151–1155, 2002a.

H. Schramm, M. Kaspereit, A. Kienle, and A. Seidel-Morgenstern. Simulated moving bed processes with cyclic modulation of the feed concentration. In *Book of abstracts of the Symposium on Preparative and Industrial Chromatography and Allied Techniques*, Heidelberg, Germany, 2002b.

A. Seidel-Morgenstern. *Mathematische Modellierung der präparativen Flüssigchromatographie*. Deutscher Universitäts–Verlag, Wiesbaden, 1995.

A. M. Spieker. *Modellierung, Simulation und Optimierung der präparativen Flüssigchromatographie*. Dissertation, Universität Stuttgart, 2000.

G. Storti, M. Mazzotti, M. Morbidelli, and S. Carra. Robust design of binary countercurrent adsorption separation processes. *American Institute of Chemical Engineers Journal (AIChE)*, 39:471–492, 1993.

A. Toumi. *Optimaler Betrieb und Regelung von Simulated–Moving–Bed–Prozessen*. Dissertation, Universität Dortmund, 2005.

A. Toumi, S. Engell, and F. Hanisch. Asynchron getaktete Gegenstromchromatographie – Prinzip und optimaler Betrieb. *Chemie Ingenieur Technik*, 75:1483–1490, 2002a.

A. Toumi, F. Hanisch, and S. Engell. Optimal operation of continuous chromatographic processes: Mathematical optimization of the VARICOL process. *Industrial and Engineering Chemistry Research*, 41:4328–4337, 2002b.

A. v. d. Schaft and H. Schumacher. *An Introduction to Hybrid Dynamical Systems*. Springer–Verlag London, 2000.

G. Zimmer, G. Dünnebier, and K.-U. Klatt. On–line parameter estimation and process monitoring of Simulated Moving Bed chromatographic separation processes. In *Proceedings of the European Control Confenrence ECC 1999*, Karlsruhe, Germany, 1999.

Appendix A

Symbols

Table A.1: Abbreviations

LTV	Linear time–varying
SCC	Simulated Counterflow Chromatography
SMB	Simulated Moving Bed
TMB	True Moving Bed
VARICOL	Variable Length Column

Table A.2: Symbols

A	Mixture component, product outlet port
$A + B$	Feed mixture, feed inlet port
B	Mixture component, product outlet port
D	Diffusion coefficient
E	Switching signal, event generator relation
F	Void fraction
G	Discrete state transition relation
H	Linear adsorption coefficient, discrete output relation
I	Injector relation
K	Switching cycle, controller
L	Column length
P	Set of operation parameters
S	Solvent, solvent inlet port
R	Recycling
T	Switching time, time delay, time constant

V	Volume
W	Frequency domain input variable
X	Set of continuous states
Y	Frequency domain output variable
a	Characteristic polynomial coefficient
b	Hill isotherm parameter, wave front parameter
c	Concentration in the mobile phase
c_1, c_2, c_3, c_4	Wave front
d	Column diameter
e	Observer output error
g	Impulse response, state–space model output function
h	Step response, characteristic polynomial coefficient
i	Integer, counter
j	SCC section (latin number)
k	Switching period, discrete time, gain
m	Partial mass, column number
$\dot{m}$	Mass flow rate
n	Integer, counter, vector dimension
p	Parameter
q	Adsorbent surface concentration, parameter
r	Purity value, parameter
s	Parameter, complex variable
t	Global time
u	Fluid flow velocity, controller variable
v	Relative fluid flow velocity
w	Vector of the fluid flow rates in the SCC columns
z	Spatial coordinate
$\boldsymbol{A}$	System matrix
$\boldsymbol{C}$	Output matrix
$\boldsymbol{G}$	Impulse response matrix
$\boldsymbol{H}$	Observer error system matrix, step response matrix
$\boldsymbol{I}$	Unity matrix
$\boldsymbol{S}$	Observability matrix
$\boldsymbol{Q}$	Transformation matrix

$\boldsymbol{c}'$	Output vector
$\boldsymbol{d}$	Continuous disturbance input
$\boldsymbol{f}$	Vector operator for the compact distributed system model state equation, vector function of the state–space model state equation
$\boldsymbol{g}$	Vector operator for the compact distributed system model output equation
$\boldsymbol{l}$	Observer gain
$\boldsymbol{u}$	Continuous input
$\boldsymbol{w}$	Set–point vector
$\boldsymbol{x}$	Distributed continuous state
$\boldsymbol{y}$	Continuous output
$\hat{\boldsymbol{x}}$	Observer state
$\hat{\boldsymbol{y}}$	Observer output
$\mathbb{R}$	Set of real numbers
$\mathbb{N}$	Set of integers
v	Vector of discrete input
w	Vector of discrete output
z	Vector of discrete state
$\mathcal{A}$	Hybrid automaton
$\mathcal{N}$	Set of discrete variables
$\mathcal{D}$	Set of discrete state vectors
$\mathcal{O}$	Set of discrete output vectors
$\mathcal{S}$	Dynamical system
$\mathcal{W}$	Wave front model

Table A.3: Special indices

B	Mixture component, observer, observer canonical form
C	Convection
D	Diffusion
Eq	Adsorption equilibrium
I	Integrator
Q	Adsorbent

b	By–product, observer
bc	Boundary condition
c	Continuous, column, continuous controller
d	Discrete, delay
dyn	Dynamical
ic	Initial condition
in	Input, inner control loop
m	Measurement, clomun number
max	Largest impurity concentration
out	Output, outer control loop
s	Liquid saturation concentration for Hill isotherms
set	Set–point
$stat$	Stationary
rel	Relative
w	Closed–loop
0	Initial
$*$	Placeholder for $S, A, A + B, B, R$

Table A.4: Greek letters

$\mathbf{\Phi}$	State transition matrix
α	Tuning factor
β	Tuning factor
ε	Adsorbent package porosity
κ	Sub–period
λ	Eigenvalue
μ	Explicit functional description, labelling function
ρ	Density
τ	Local time
$\boldsymbol{\zeta}$	Lumped distributed and continuous state

Table A.5: Special notations

F'	Weighting factor
T_S	Switching time of S
c_b	By–product concentration
dc_m	Concentration measurement error
dV	Balance volume
n_c	Number of SCC columns
$\bar{n}_c$	Mean number of columns
t_s	Sampling time
z_0	Position, wave front parameter
ΔT	Relative switching time
δc	Observation error
$\delta \bar{c}$	Mean observation error
δT	Sub–period duration

Appendix B

Example separation process data

For the numerical simulation of the separation processes the cubic Hill isotherm is used to describe the adsorption equilibrium. The function is given by

$$q_*(c_*) = \frac{q_{s,*}}{3} \frac{b_1 \, c_* + 2 \, b_2 \, c_*^2 + 3 \, b_3 \, c_*^3}{1 + b_1 \, c_* + b_2 \, c_*^2 + b_3 \, c_*^3} .$$

The parameters of the considered separation plant setup as well as the parameters of the isotherm and the considered nominal operation point are given in the tables below.

Data of the Ibuprofen separation.

Table B.1: Plant data

Plant data				
L_c	=	0,13	Column length	[m]
D_c	=	0,03	Column diameter	[m]
V_d	=	$1 \cdot 10^{-5}$	Extra–column volume of one column interconnection	$[\mathrm{m}^3]$
Adsorbent package parameters				
ε	=	0,728	Adsorbent package porosity	[1]
D	=	$1 \cdot 10^{-5}$	Diffusion coefficient	$\left[\frac{\mathrm{m}^2}{\mathrm{s}}\right]$

Table B.2: Ibuprofen isotherm parameters

Isotherm parameters of R-Ibuprofen (component A)

$q_{s,A}$	=	136,76	$\left[\frac{\mathrm{kg}}{\mathrm{m}^3}\right]$
$b_{1,A}$	=	0,2135	$\left[\frac{\mathrm{m}^3}{\mathrm{kg}}\right]$
$b_{2,A}$	=	0,0156	$\left[\frac{\mathrm{m}^3}{\mathrm{kg}}^2\right]$
$b_{3,A}$	=	0,0012	$\left[\frac{\mathrm{m}^3}{\mathrm{kg}}^3\right]$

Isotherm parameters of S-Ibuprofen (component B)

$q_{s,B}$	=	146,078	$\left[\frac{\mathrm{kg}}{\mathrm{m}^3}\right]$
$b_{1,B}$	=	0,1672	$\left[\frac{\mathrm{m}^3}{\mathrm{kg}}\right]$
$b_{2,B}$	=	0,0094	$\left[\frac{\mathrm{m}^3}{\mathrm{kg}}^2\right]$
$b_{3,B}$	=	0,0006	$\left[\frac{\mathrm{m}^3}{\mathrm{kg}}^3\right]$

Table B.3: Plant and operation point parameters

Nominal operation point

T_S	=	120	Switching time	[s]
$c_{A+B,A}$	=	1	Feed inlet concentration of component A	$\left[\frac{\mathrm{kg}}{\mathrm{m}^3}\right]$
$c_{A+B,B}$	=	1	Feed inlet concentration of component B	$\left[\frac{\mathrm{kg}}{\mathrm{m}^3}\right]$
$\dot{m}_I$	=	$2{,}11 \cdot 10^{-3}$	Fluid mass flow rate section I	$\left[\frac{\mathrm{kg}}{\mathrm{s}}\right]$
$\dot{m}_{II}$	=	$1{,}66 \cdot 10^{-3}$	Fluid mass flow rate section II	$\left[\frac{\mathrm{kg}}{\mathrm{s}}\right]$
$\dot{m}_{III}$	=	$1{,}74 \cdot 10^{-3}$	Fluid mass flow rate section III	$\left[\frac{\mathrm{kg}}{\mathrm{s}}\right]$
$\dot{m}_{IV}$	=	$1{,}38 \cdot 10^{-3}$	Fluid mass flow rate section IV	$\left[\frac{\mathrm{kg}}{\mathrm{s}}\right]$
p_S	=	169	Pressure at the solvent inlet port S	[bar]
Δp_c	=	3,13	Pressure drop over one column S	[bar]

Data of the Tocopherol separation.

Table B.4: Plant data

Plant data				
L_c	=	0,075	Column length	[m]
D_c	=	0,03	Column diameter	[m]
V_d	=	$1 \cdot 10^{-5}$	Extra–column volume of one column interconnection	$[\mathrm{m}^3]$
Adsorbent package parameters				
ε	=	0,755	Adsorbent package porosity	[1]
D	=	$4 \cdot 10^{-6}$	Diffusion coefficient	$\left[\frac{\mathrm{m}^2}{\mathrm{s}}\right]$

Table B.5: Tocopherol isotherm parameters

Isotherm parameters of α-Tocopherol (component A)			
$q_{s,A}$	=	4314	$\left[\frac{\mathrm{kg}}{\mathrm{m}^3}\right]$
$b_{1,A}$	=	0,0078	$\left[\frac{\mathrm{m}^3}{\mathrm{kg}}\right]$
$b_{2,A}$	=	0,0002	$\left[\frac{\mathrm{m}^3}{\mathrm{kg}}^2\right]$
$b_{3,A}$	=	0,00002	$\left[\frac{\mathrm{m}^3}{\mathrm{kg}}^3\right]$
Isotherm parameters of δ-Tocopherol (component B)			
$q_{s,B}$	=	6860,1	$\left[\frac{\mathrm{kg}}{\mathrm{m}^3}\right]$
$b_{1,B}$	=	0,0086	$\left[\frac{\mathrm{m}^3}{\mathrm{kg}}\right]$
$b_{2,B}$	=	0,0003	$\left[\frac{\mathrm{m}^3}{\mathrm{kg}}^2\right]$
$b_{3,B}$	=	0,00003	$\left[\frac{\mathrm{m}^3}{\mathrm{kg}}^3\right]$

Table B.6: Plant and operation point parameters

Nominal operation point

T_S	=	240	Switching time	[s]
$c_{A+B,A}$	=	1	Feed inlet concentration of component A	$\left[\frac{\mathrm{kg}}{\mathrm{m}^3}\right]$
$c_{A+B,B}$	=	1	Feed inlet concentration of component B	$\left[\frac{\mathrm{kg}}{\mathrm{m}^3}\right]$
$\dot{m}_I$	=	$1{,}13 \cdot 10^{-3}$	Fluid mass flow rate section I	$\left[\frac{\mathrm{kg}}{\mathrm{s}}\right]$
$\dot{m}_{II}$	=	$8{,}58 \cdot 10^{-4}$	Fluid mass flow rate section II	$\left[\frac{\mathrm{kg}}{\mathrm{s}}\right]$
$\dot{m}_{III}$	=	$8{,}89 \cdot 10^{-4}$	Fluid mass flow rate section III	$\left[\frac{\mathrm{kg}}{\mathrm{s}}\right]$
$\dot{m}_{IV}$	=	$6{,}15 \cdot 10^{-4}$	Fluid mass flow rate section IV	$\left[\frac{\mathrm{kg}}{\mathrm{s}}\right]$
p_S	=	212	Pressure at the solvent inlet port S	[bar]
Δp_c	=	2,5	Pressure drop over one column S	[bar]

Appendix C

Special derivations and experimental data

C.1 Initial conditions and stationary solution of the True Moving Bed model

The general solution (3.35) of the TMB model (3.31) has to fulfil the initial conditions (3.32) for each TMB section. Therefore, for $t = 0$ the general solution (3.35) is expressed in terms of the initial conditions (3.32), for which the expressions C.1 are obtained.

The solution of (C.1) (obtained by the application numerical methods) yields the parameters $p_{1,*,j,i}$, $q_{1,*,j,i}$, $p_{2,*,j,i}$ and $q_{2,*,j,i}$. The solution of the system made up by Equations (3.37) and (3.38) yields an explicit expression for the parameters $r_{*,j}$ and $s_{*,j}$ given by Equations (C.2) through (C.5).

Equations (3.36) and (C.2) through (C.5) are the analytic solution of the stationary TMB process for the component $*$.

$$\begin{aligned}
I:\ z\in[0,1] \quad c_{A,I}(0,z) &= \sum_{i=1}^{\infty}\left(p_{1,A,I,i}\cdot \mathrm{e}^{q_{1,A,I,i}\,z}+p_{2,A,I,i}\cdot \mathrm{e}^{q_{2,A,I,i}\,z}\right)\\
&+ r_{A,I}\cdot \mathrm{e}^{\frac{v_{A,I}}{D}\,z}+s_{A,I}\,,\\
c_{B,I}(0,z) &= \sum_{i=1}^{\infty}\left(p_{1,B,I,i}\cdot \mathrm{e}^{q_{1,B,I,i}\,z}+p_{2,B,I,i}\cdot \mathrm{e}^{q_{2,B,I,i}\,z}\right)\\
&+ r_{B,I}\cdot \mathrm{e}^{\frac{v_{B,I}}{D}\,z}+s_{B,I}\,,\\
II:\ z\in[1,2] \quad c_{A,II}(0,z) &= \sum_{i=1}^{\infty}\left(p_{1,A,II,i}\cdot \mathrm{e}^{q_{1,A,II,i}\,(z-1)}+p_{2,A,II,i}\cdot \mathrm{e}^{q_{2,A,II,i}\,(z-1)}\right)\\
&+ r_{A,II}\cdot \mathrm{e}^{\frac{v_{A,II}}{D}\,(z-1)}+s_{A,II}\,,\\
c_{B,II}(0,z) &= \sum_{i=1}^{\infty}\left(p_{1,B,II,i}\cdot \mathrm{e}^{q_{1,B,II,i}\,(z-1)}+p_{2,B,II,i}\cdot \mathrm{e}^{q_{2,B,II,i}\,(z-1)}\right)\\
&+ r_{B,II}\cdot \mathrm{e}^{\frac{v_{B,II}}{D}\,(z-1)}+s_{B,II}\,,\\
III:\ z\in[2,3] \quad c_{A,III}(0,z) &= \sum_{i=1}^{\infty}\left(p_{1,A,III,i}\cdot \mathrm{e}^{q_{1,A,III,i}\,(z-2)}+p_{2,A,III,i}\cdot \mathrm{e}^{q_{2,A,III,i}\,(z-2)}\right)\\
&+ r_{A,III}\cdot \mathrm{e}^{\frac{v_{A,III}}{D}\,(z-2)}+s_{A,III}\,,\\
c_{B,III}(0,z) &= \sum_{i=1}^{\infty}\left(p_{1,B,III,i}\cdot \mathrm{e}^{q_{1,B,III,i}\,(z-2)}+p_{2,B,III,i}\cdot \mathrm{e}^{q_{2,B,III,i}\,(z-2)}\right)\\
&+ r_{B,III}\cdot \mathrm{e}^{\frac{v_{B,III}}{D}\,(z-2)}+s_{B,III}\,,\\
IV:\ z\in[3,4] \quad c_{A,IV}(0,z) &= \sum_{i=1}^{\infty}\left(p_{1,A,IV,i}\cdot \mathrm{e}^{q_{1,A,IV,i}\,(z-3)}+p_{2,A,IV,i}\cdot \mathrm{e}^{q_{2,A,IV,i}\,(z-3)}\right)\\
&+ r_{A,IV}\cdot \mathrm{e}^{\frac{v_{A,IV}}{D}\,(z-3)}+s_{A,IV}\,,\\
c_{B,IV}(0,z) &= \sum_{i=1}^{\infty}\left(p_{1,B,IV,i}\cdot \mathrm{e}^{q_{1,B,IV,i}\,(z-3)}+p_{2,B,IV,i}\cdot \mathrm{e}^{q_{2,B,IV,i}\,(z-3)}\right)\\
&+ r_{B,IV}\cdot \mathrm{e}^{\frac{v_{B,IV}}{D}\,(z-3)}+s_{B,IV}\,,
\end{aligned} \tag{C.1}$$

with

$$\begin{aligned}
q_{1,A,j,i} &= \frac{v_{A,j}}{2D}+\sqrt{\frac{v_{A,j}^2}{4D^2}+\frac{\lambda_{A,j,i}}{F'_A\,D}}\\
q_{2,A,j,i} &= \frac{v_{A,j}}{2D}-\sqrt{\frac{v_{A,j}^2}{4D^2}+\frac{\lambda_{A,j,i}}{F'_A\,D}}\\
q_{1,B,j,i} &= \frac{v_{B,j}}{2D}+\sqrt{\frac{v_{B,j}^2}{4D^2}+\frac{\lambda_{B,j,i}}{F'_B\,D}}\\
q_{2,B,j,i} &= \frac{v_{B,j}}{2D}-\sqrt{\frac{v_{B,j}^2}{4D^2}+\frac{\lambda_{B,j,i}}{F'_B\,D}}\,.
\end{aligned} \qquad j=I,II,III,IV$$

$$\begin{aligned}
r_{*,I} &= \frac{(v_{*,III}-v_{*,II})\,d_*\,c_{*,in}}{a_*\,d_*-b_*\,c_*}\\
s_{*,I} &= -\frac{(v_{*,III}-v_{*,II})\,c_*\,c_{*,in}}{a_*\,d_*-b_*\,c_*}\,,
\end{aligned} \tag{C.2}$$

$$
\begin{aligned}
r_{*,II} &= \frac{v_{*,I}}{v_{*,II}} \alpha_{*,I} r_{*,I} \\
r_{*,III} &= -\frac{\beta_{*,I}}{\beta_{2,*}} r_{*,I} \\
r_{*,IV} &= -\frac{v_{*,III}}{v_{*,IV}} \alpha_{*,III} \frac{\beta_{1,*}}{\beta_{2,*}} r_{*,I} ,
\end{aligned}
\tag{C.3}
$$

$$
\begin{aligned}
s_{*,II} &= r_{*,I} \alpha_{*,I} \left(1 - \frac{v_{*,I}}{v_{*,II}}\right) + s_{*,I} \\
s_{*,III} &= r_{*,I} \left(1 - \beta_{1,*} + \frac{\beta_{1,*}}{\beta_{2,*}}\right) + s_{*,I} \\
s_{*,IV} &= r_{*,I} \left(1 + \alpha_{*,III} \alpha_{*,IV} \frac{v_{*,III}}{v_{*,IV}} \frac{\beta_{1,*}}{\beta_{2,*}}\right) + s_{*,I} .
\end{aligned}
\tag{C.4}
$$

The following abbreviations are used:

$$
\begin{aligned}
\alpha_{*,j} &= \mathrm{e}^{\frac{v_{*,j}}{D}} , \quad j = I, II, III, IV \\
\beta_{1,*} &= 1 - \alpha_{*,I} + \frac{v_{*,I}}{v_{*,II}} \alpha_{*,I} (1 - \alpha_{*,II}) \\
\beta_{2,*} &= 1 - \alpha_{*,III} + \frac{v_{*,III}}{v_{*,IV}} \alpha_{*,III} (1 - \alpha_{*,IV}) , \\
a_* &= v_{*,I} \alpha_{*,I} \alpha_{*,II} + v_{*,II} (\beta_{1,*} - 1) + v_{*,III} \left(1 - \beta_{1,*} + \frac{\beta_{1,*}}{\beta_{2,*}}\right) \\
b_* &= v_{*,III} - v_{*,II} \\
c_* &= v_{*,III} \alpha_{*,III} \alpha_{*,IV} \frac{\beta_{1,*}}{\beta_{2,*}} + v_{*,IV} \\
d_* &= v_{*,III} - v_{*,I} .
\end{aligned}
\tag{C.5}
$$

C.2 Example of a stationary True Moving Bed process

For the calculation of the stationary TMB concentration profile, the following parameters are used:

$$
\begin{gathered}
v_{A,I} = 0{,}0011 , \quad v_{A,II} = -0{,}0007 , \quad v_{A,III} = -0{,}0006 , \quad v_{A,IV} = -0{,}0027 , \\
v_{B,I} = 0{,}0027 , \quad v_{B,II} = 0{,}0008 , \quad v_{B,III} = 0{,}0009 , \quad v_{B,IV} = -0{,}0011 , \\
D = 1 \cdot 10^{-4} , \quad c_{in,A} = c_{in,B} = 1 .
\end{gathered}
$$

Applying the parameters to Equation (3.36) yields

$$\begin{aligned}
I: \quad z \in [0,1] \quad & c_{stat,A,I}(z) = 7{,}96 \cdot 10^{-7} \cdot \mathrm{e}^{11{,}5 \cdot z} - 5{,}56 \cdot 10^{-7} \\
& c_{stat,B,I}(z) = 3{,}27 \cdot 10^{-17} \cdot \mathrm{e}^{26{,}9 \cdot z} + 2{,}94 \cdot 10^{-7} \\
II: \quad z \in [1,2] \quad & c_{stat,A,II}(z) = -1{,}20 \cdot 10^{-1} \cdot \mathrm{e}^{-7{,}37 \cdot (z-1)} + 1{,}96 \cdot 10^{-1} \\
& c_{stat,B,II}(z) = 5{,}09 \cdot 10^{-5} \cdot \mathrm{e}^{8{,}01 \cdot (z-1)} - 3{,}54 \cdot 10^{-5} \\
III: \quad z \in [2,3] \quad & c_{stat,A,III}(z) = 1{,}97 \cdot 10^{-1} \cdot \mathrm{e}^{-5{,}92 \cdot (z-2)} - 4{,}08 \cdot 10^{-4} \\
& c_{stat,B,III}(z) = -6{,}46 \cdot 10^{-6} \cdot \mathrm{e}^{9{,}46 \cdot (z-2)} + 1{,}53 \cdot 10^{-1} \\
IV: \quad z \in [3,4] \quad & c_{stat,A,IV}(z) = 1{,}17 \cdot 10^{-4} \cdot \mathrm{e}^{-26{,}5 \cdot (z-3)} + 2{,}40 \cdot 10^{-7} \\
& c_{stat,B,IV}(z) = 7{,}03 \cdot 10^{-2} \cdot \mathrm{e}^{-11{,}2 \cdot (z-3)} - 7{,}08 \cdot 10^{-7} \ .
\end{aligned} \tag{C.6}$$

This result is shown in Figure 3.9.

C.3 Mean section length of the VARICOL process

The mean section length $\bar{n}_{c,j}$ of a VARICOL process is determined by the switching pattern. This appendix shows the derivation of a general expression, which allows to determine the mean section length in dependence upon the switching pattern. The solution is derived by applying Equation (2.23) for a given initial column configuration

$$(n_{c,I}(\kappa_0)/n_{c,II}(\kappa_0)/n_{c,III}(\kappa_0)/n_{c,IV}(\kappa_0))$$

and for given switching times T_S, ΔT_A, ΔT_{A+B} and ΔT_B of the considered switching period k_S. Thereby, the the number of columns $n_{c,j}(\kappa)$ and the duration δT_κ of the sub–periods of k_S is determined.

Figure C.1 shows the number of columns in the VARICOL section I during the switching period k_S. The Figure shows that there are $n_{c,I} = n_{c,0,I}$ columns in section I for the duration of $\delta T_1 = \Delta T_A$, and $n_{c,I} = n_{c,0,I} + 1$ columns for the duration of $\delta T_2 + \delta T_3 + \delta T_4 = T_S - \Delta T_A$. Hence, the mean section length $\bar{n}_{c,I}$ is

$$\bar{n}_{c,I} = n_{c,0,I} + \frac{T_S - \Delta T_A}{T_S} \ . \tag{C.7}$$

If the ports S and A are switched synchronously, ΔT_A has to be set to $\Delta T_A = T_S$.

Figure C.2 shows the number of columns in the VARICOL section II during a switching period k_S. Two cases are possible: either $\Delta T_A \leq \Delta T_{A+B}$, or $\Delta T_A \geq \Delta T_{A+B}$ holds. For both cases the mean section length is

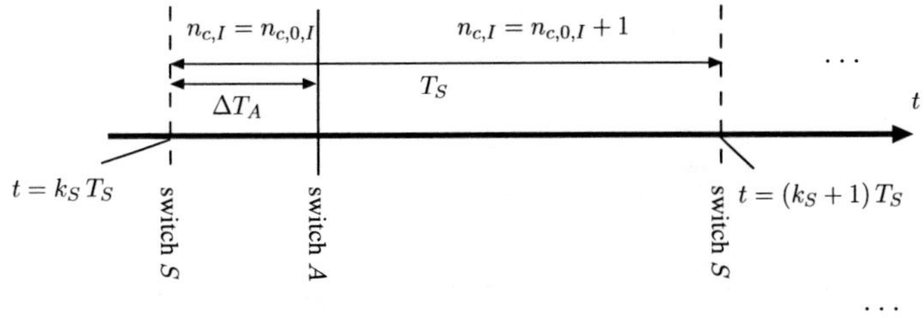

Figure C.1: Number of columns in section I during a switching period k_S

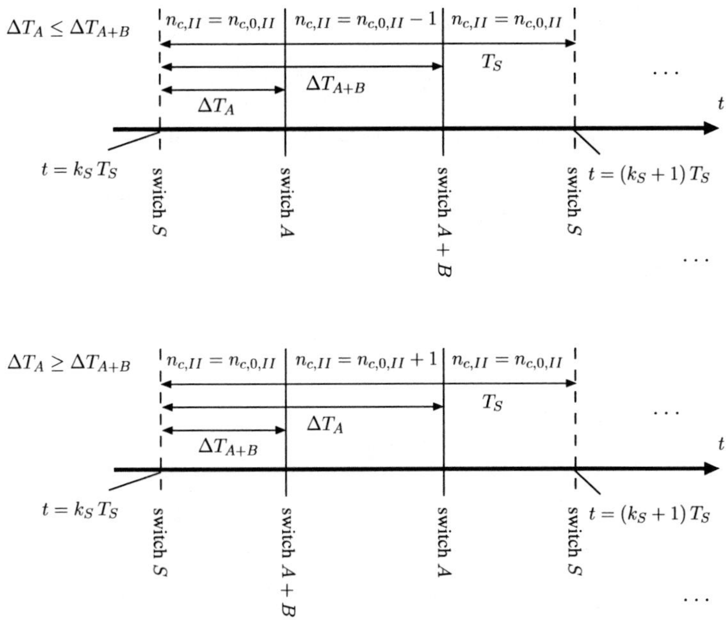

Figure C.2: Number of columns in section II during a switching period k_S

$$\bar{n}_{c,II} = n_{c,0,II} + \frac{\Delta T_A - \Delta T_{A+B}}{T_S}. \tag{C.8}$$

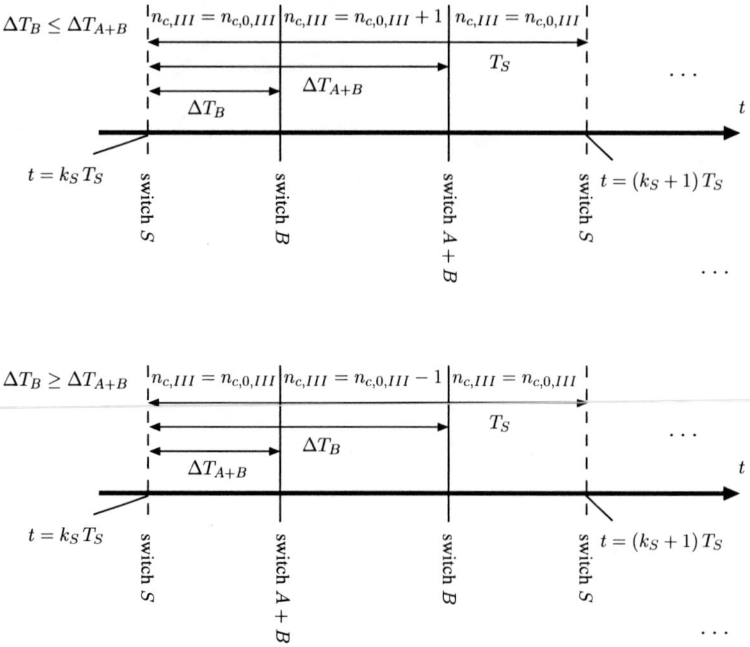

Figure C.3: Number of columns in section III during a switching period k_S

The number of columns in section III during a VARICOL switching period are shown in Figure C.3. Like above, two cases occur: either $\Delta T_B \leq \Delta T_{A+B}$, or $\Delta T_B \geq \Delta T_{A+B}$ holds. For both cases, the mean section length $\bar{n}_{c,III}$ is

$$\bar{n}_{c,III} = n_{c,0,III} + \frac{\Delta T_{A+B} - \Delta T_B}{T_S}. \tag{C.9}$$

For section IV, the number of columns during one switching period k_S is represented in Figure C.4. As in the case of section I, only one case occurs. Here, the mean section length $\bar{n}_{c,IV}$ is

$$\bar{n}_{c,IV} = n_{c,0,IV} + \frac{\Delta T_B - T_S}{T_S}. \tag{C.10}$$

The Equations (C.7) to (C.10) show, how the mean section lengths are determined by the switching pattern, i.e. by the switching time T_S, the relative switching times ΔT_A, ΔT_{A+B} and ΔT_B, and the initial column configuration $(n_{c,0,I}/n_{c,0,II}/n_{c,0,III}/n_{c,0,IV})$.

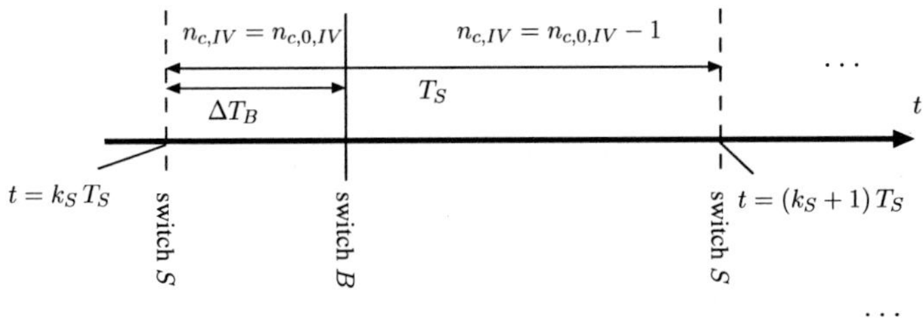

Figure C.4: Number of columns in section IV during a switching period k_S

C.4 Number of port position combinations

For the derivation of an explicit expression to determine the number of port position combinations of the VARICOL in dependence upon n_c, all possible port positions on a circle of separation columns, which fulfil the conditions (3.53) and (3.54), are counted. The solution is determined considering an example of a four column VARICOL process. The result is generalised for processes with n_c columns. The principle of counting the possible port positions is shown in Figure C.5 and is performed as follows:

1. The position of the solvent inlet is set to $z_S = 0$.

2. All other ports are positioned as close as possible to z_S while regarding Equations (3.53) and (3.54):

$$z_A = 0\,, \quad z_{A+B} = 0\,, \quad z_B = 1\,.$$

   This initial setup is shown in Figure C.5 for $\mathsf{w} = (\ 0 \quad 0 \quad 0 \quad 1\ )'$.

3. The numbers of possible positions of the last port, z_B, is evaluated. The following output variables are obtained including the initial output:

$$\mathsf{w} = (\ 0\ \ 0\ \ 0\ \ 1\ )'$$
$$\mathsf{w} = (\ 0\ \ 0\ \ 0\ \ 2\ )'$$
$$\mathsf{w} = (\ 0\ \ 0\ \ 0\ \ 3\ )'$$
$$\mathsf{w} = (\ 0\ \ 0\ \ 0\ \ 4\ )'.$$

In Figure C.5, these output variables are represented by the numbers 1. to 4.

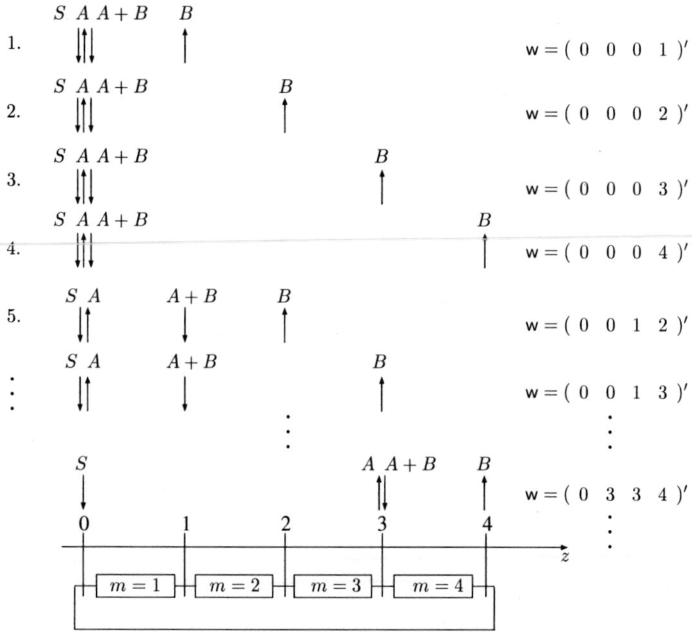

Figure C.5: Analysis of the inlet and outlet port positions

4. The next step of the analysis considers the port position 5. in Figure C.5. This port position is obtained considering the initial port position (1.) and switching the port $A + B$ by one column length. Because there must always be at least one column in the section III, the port B has also to be switched. Starting from this port position combination, all possible positions of z_B are counted. Three possibilities with the following output variables are obtained:

$$\mathsf{w} = (\ 0\ \ 0\ \ 1\ \ 2\ )'$$
$$\mathsf{w} = (\ 0\ \ 0\ \ 1\ \ 3\ )'$$
$$\mathsf{w} = (\ 0\ \ 0\ \ 1\ \ 4\ )'.$$

Now, the port position z_A can also be moved form position $z_A = 0$ to position $z_A = 1$. Evaluating the possible positions of the port B yields

$$\mathsf{w} = (\ 0\ \ 1\ \ 1\ \ 2\ )'$$
$$\mathsf{w} = (\ 0\ \ 1\ \ 1\ \ 3\ )'$$
$$\mathsf{w} = (\ 0\ \ 1\ \ 1\ \ 4\ )'.$$

5. Next, the position z_{A+B} is again switched by one column to $z_{A+B} = 2$ and z_A is positioned back to $z_A = 0$. Because of condition (3.54) the position of the port B must be $z_B = 3$. Exploring the possible position of z_B yields

$$\mathsf{w} = (\ 0\ \ 0\ \ 2\ \ 3\ )'$$
$$\mathsf{w} = (\ 0\ \ 0\ \ 2\ \ 4\ )'.$$

Now all possible combinations for a positioning of z_B are counted for $z_A = 1$ and $z_A = 2$. The results reveal the following output variables:

$$\mathsf{w} = (\ 0\ \ 1\ \ 2\ \ 3\ )'$$
$$\mathsf{w} = (\ 0\ \ 1\ \ 2\ \ 4\ )'$$
$$\mathsf{w} = (\ 0\ \ 2\ \ 2\ \ 3\ )'$$
$$\mathsf{w} = (\ 0\ \ 2\ \ 2\ \ 4\ )'.$$

6. Again, shifting z_{A+B} by one column to $z_{A+B} = 3$, the position z_B can only be $z_B = 4$ and for $z_A = 0$ the output

$$\mathsf{w} = (\ 0\ \ 0\ \ 3\ \ 4\ )'$$

is obtained. Now considering all possible port positions of the port A for $z_A = 1$, $z_A = 2$ and $z_A = 3$ reveals

$$\mathsf{w} = (\ 0\ \ 1\ \ 3\ \ 4\ )'$$
$$\mathsf{w} = (\ 0\ \ 2\ \ 3\ \ 4\ )'$$
$$\mathsf{w} = (\ 0\ \ 3\ \ 3\ \ 4\ )'.$$

7. Counting all port positions reveals 20 possibilities for a VARICOL setup with $n_c = 4$ columns and for the fixed position $z_S = 0$. Because S can take $n_c = 4$ positions in the whole, the total number of possible port position combinations, which fulfil the requirements (3.53) and (3.54), for the four column VARICOL is

$$n_w = n_c \cdot 20 = 80\,.$$

Applying this principle of counting the port position combinations to a VARICOL with n_c columns, the following explicit expression of n_w in dependence upon n_c is obtained for a fixed spatial coordinate system:

$$n_w = \sum_{i=1}^{n_c-1}\left(\sum_{j=0}^{n_c-i} j+1\right) n_c = \left(\sum_{i=1}^{n_c}(n_c+1)\,i\;-\;i^2\right) n_c\,.$$

For a theoretical VARICOL setup with three columns, this reveals

$$n_w(n_c = 3) = 30$$

and for $n_c = 5$ and $n_c = 6$

$$\begin{aligned} n_w(n_c = 5) &= 175 \\ n_w(n_c = 5) &= 336\,. \end{aligned}$$

C.5 Experimental step responses

An experiment for the verification of the dynamical behaviour of SMB processes was performed with the SFC SMB plant described in Section 2.7. The separation of α– and δ–Tocopherol was performed to which a simultaneous step change of all internal fluid flow rates was applied. Concentration measurements as described in Section 2.7 were performed with $\tau_m = \frac{1}{2}T_S$. The chromatogram of the whole experiment was recorded (left plot of Figure C.6). After the 7th cycle the simultaneous step input was applied. For the evaluation of the results, the concentration measurements of each cycle were analysed separately. The right plot in Figure C.6 shows a close–up of the 7th cycle. To each peak of the UV signal, a component and a measurement position is assigned. Based on this method, the concentration profiles shown in Figures C.7 and C.8 are determined. For each of the measuremet positions the evolutions of the respective concentrations are determined from the concentration profiles and plotted over K. The results for the wave front concentration evolution are shown in Section 5.5.3.

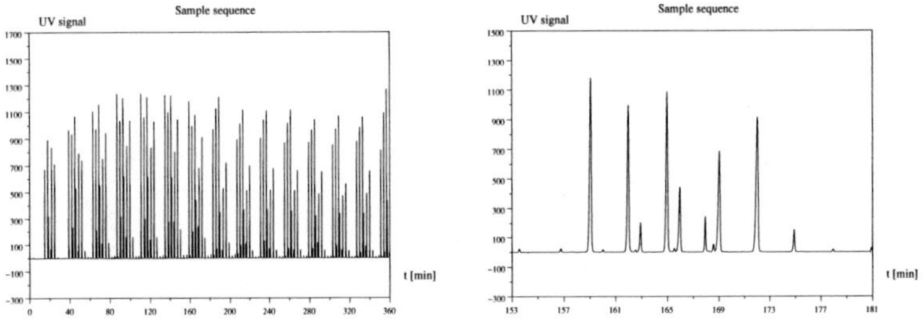

Figure C.6: Sample sequence of the experiment and of the cycle $K = 7$

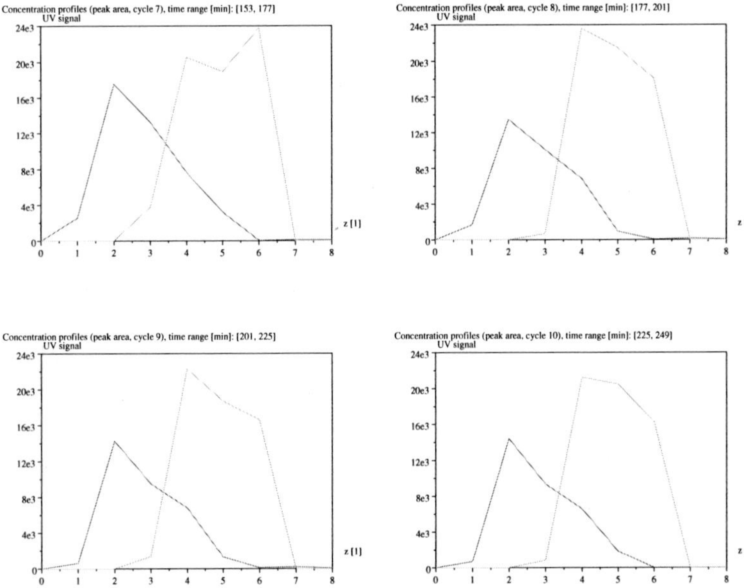

Figure C.7: Cycles 7 to 10

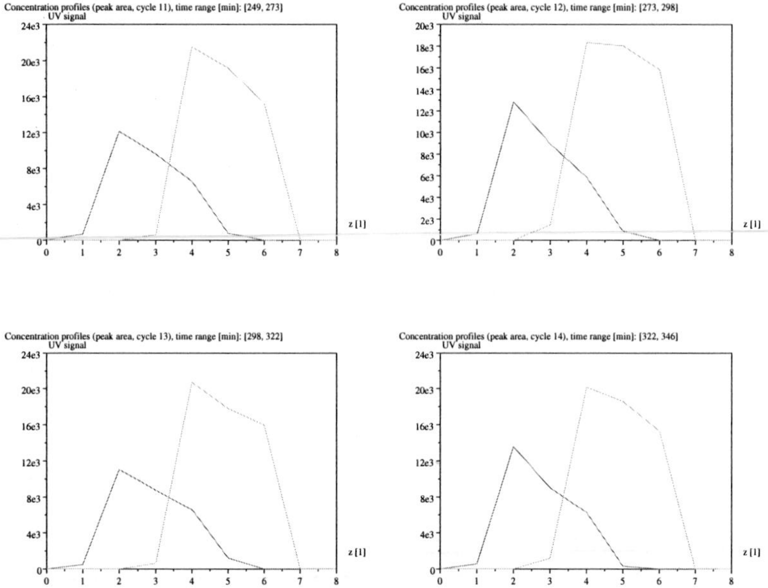

Figure C.8: Cycles 11 to 14

Curriculum vitae

Tobias Kleinert, born 1971, December 16th, in Herdecke (Germany)

since 06/2005	Process engineer BASF AG Ludwigshafen (Germany) Service Center Automation Technology
03/2002–05/2005	Scientific coworker Ruhr-Universität Bochum (Germany) Automatisierungstechnik und Prozessinformatik
05/2000–02/2002	Scientific coworker Technische Universität Hamburg–Harburg (Germany) Arbeitsbereich Regelungstechnik
01/2000–04/2000	Scientific coworker Rheinisch–Westfälische Technische Hochschule Aachen (Germany) Institut für Regelungstechnik
05/1998–09/1998	Student project ADERSA, Palaiseau (France)
01/1998–02/1998	Student project Sulzer Infra Lab, Winterthur (Switzerland)
04/1995–12/1999	Diploma studies of Thermal Engineering Rheinisch–Westfälische Technische Hochschule Aachen (Germany) honoured with the Friedrich–Wilhelm–Preis, RWTH Aachen
10/1992–03/1995	Studies of Mechanical Engineering (Vordiplom) Universität Dortmund (Germany)